CLASSICAL MYTHOLOGY
OF THE
CONSTELLATIONS

CLASSICAL MYTHOLOGY
OF THE
CONSTELLATIONS

TIMELESS TALES OF THE
STARRY NIGHT SKY

ANNETTE GIESECKE

ILLUSTRATIONS BY
JIM TIERNEY

BLACK DOG
& LEVENTHAL
PUBLISHERS
NEW YORK

Copyright © 2025 by Annette Giesecke
Interior and jacket illustrations copyright © 2025 by Jim Tierney

Jacket copyright © 2025 by Hachette Book Group, Inc.

Hachette Book Group supports the right to free expression and the value of copyright. The purpose of copyright is to encourage writers and artists to produce the creative works that enrich our culture.

The scanning, uploading, and distribution of this book without permission is a theft of the author's intellectual property. If you would like permission to use material from the book (other than for review purposes), please contact permissions@hbgusa.com. Thank you for your support of the author's rights.

Black Dog & Leventhal Publishers
Hachette Book Group
1290 Avenue of the Americas, New York, NY 10104
www.blackdogandleventhal.com
BlackDogandLeventhal @BDLev

First Edition: August 2025

Published by Black Dog & Leventhal Publishers, an imprint of Hachette Book Group, Inc. The Black Dog & Leventhal Publishers name and logo are trademarks of Hachette Book Group, Inc.

The Hachette Speakers Bureau provides a wide range of authors for speaking events. To find out more, go to hachettespeakersbureau.com or email HachetteSpeakers@hbgusa.com.

Black Dog & Leventhal books may be purchased in bulk for business, educational, or promotional use. For more information, please contact your local bookseller or the Hachette Book Group Special Markets Department at Special.Markets@hbgusa.com.

The publisher is not responsible for websites (or their content) that are not owned by the publisher.

Print book jacket and interior design by Katie Benezra

Library of Congress Cataloging-in-Publication Data has been applied for.

ISBNs: 978-0-7624-8851-3 (hardcover); 978-0-7624-8852-0 (ebook)

Printed in China

1010

10 9 8 7 6 5 4 3 2 1

CONTENTS

Preface xiii

INTRODUCTION

Stars and Their Myths 3
Ptolemy and His Constellations 6
Reading the Heavens 9
Further Notes for the Reader 11

PTOLEMY'S CONSTELLATIONS
NORTHERN CELESTIAL QUADRANT I

Andromeda, *Andromeda* 19
Aries, *the Ram* 21
Auriga, *the Charioteer* 24
Cassiopeia, *Cassiopeia* 29
Orion, *Orion* 31
Perseus, *Perseus* 35
Pisces, *the Fishes* 40
Taurus, *the Bull* 43
Triangulum, *the Triangle* 46

NORTHERN CELESTIAL QUADRANT II

Cancer, *the Crab* 53
Canis Minor, *the Lesser Dog* 54
Gemini, *the Twins* 57
Leo, *the Lion* 60
Ursa Major, *the Greater Bear* 63

NORTHERN CELESTIAL QUADRANT III

Boötes, *the Herdsman* 69
Corona Borealis, *the Northern Crown* 71
Draco, *the Dragon* 74
Hercules, *Hercules* 77
Ursa Minor, *the Lesser Bear* 86

NORTHERN CELESTIAL QUADRANT IV

Aquila, *the Eagle* 93
Cepheus, *Cepheus* 95
Cygnus, *the Swan* 97
Delphinus, *the Dolphin* 100
Equuleus, *the Foal* 103
Lyra, *the Lyre* 105
Pegasus, *Pegasus* 109
Sagitta, *the Arrow* 112

SOUTHERN CELESTIAL QUADRANT I

Cetus, *the Sea Monster* 119
Eridanus, *the River* 120
Lepus, *the Hare* 123

SOUTHERN CELESTIAL QUADRANT II

Argo Navis, *the Ship* Argo 129
Canis Major, *the Greater Dog* 133
Crater, *the Cup* 137
Hydra, *the Water Snake* 140

SOUTHERN CELESTIAL QUADRANT III

Ara, *the Altar* 147
Centaurus, *the Centaur* 149
Corvus, *the Crow* 152
Libra, *the Scales* 155
Lupus, *the Wolf* 158
Ophiuchus, *the Serpent-Holder* 161
Scorpius, *the Scorpion* 164
Serpens, *the Snake* 166
Virgo, *the Virgin* 169

SOUTHERN CELESTIAL QUADRANT IV

Aquarius, *the Water-Bearer* 177
Capricornus, *the Sea-Goat* 179
Corona Australis, *the Southern Crown* 182
Piscis Austrinus, *the Southern Fish* 184
Sagittarius, *the Archer* 187

CONSTELLATIONS AFTER PTOLEMY
PLANCIUS'S CONSTELLATIONS

Camelopardalis, *the Giraffe* 196
Columba, *the Dove* 196
Crux, *the Southern Cross* 197
Monoceros, *the Unicorn* 197

PLANCIUS'S COLLABORATIVE CONSTELLATIONS

Apus, *the Bird of Paradise* 198
Chamaeleon, *the Chameleon* 199
Dorado, *the Golden Fish* 199
Grus, *the Crane* 200
Hydrus, *the Male Water Snake* 200
Indus, *the Indian* 201
Musca, *the Fly* 201
Pavo, *the Peacock* 202
Phoenix, *the Phoenix* 202
Triangulum Australe, *the Southern Triangle* 203
Tucana, *the Toucan* 203
Volans, *the Flying Fish* 204

HEVELIUS'S CONSTELLATIONS

Canes Venatici, *the Hunting Dogs* 205
Lacerta, *the Lizard* 205
Leo Minor, *the Lesser Lion* 206
Lynx, *the Lynx* 206
Scutum, *the Shield* 206
Sextans, *the Sextant* 207
Vulpecula, *the Fox* 207

DE LACAILLE'S CONSTELLATIONS

Antlia, *the Air Pump* 208
Caelum, *the Chisel* 208
Carina, *the Keel* 209
Circinus, *the Drawing Compass* 209
Fornax, *the Furnace* 210
Horologium, *the Pendulum Clock* 210
Mensa, *the Table Mountain* 211

Microscopium, *the Microscope* 211
Norma, *the Set Square* 212
Octans, *the Octant* 212
Pictor, *the Painter's Easel* 213
Puppis, *the Stern* 213
Pyxis, *the Mariner's Compass* 214
Reticulum, *the Reticle* 214
Sculptor, *the Sculptor* 215
Telescopium, *the Telescope* 215
Vela, *the Sails* 216

VOPEL'S CONSTELLATION

Coma Berenices, *the Hair of Berenice* 217

APPENDICES

Constellations by the Seasons:
A Summary of Optimal Visibility by Astronomical Season 220
Ptolemy's 48 Constellations with Their Original Greek Names
(Literally Translated) 223
The Principal Gods of the Greeks and Their Roman Equivalents 224
Glossary of Ancient Sources 225

Modern Sources 228
Acknowledgments 229
Index 230

PREFACE

This book is a guide for stargazers who want to know the constellations' timeless, wondrous stories. Whether you're outside and looking up on a clear, starry night or indoors sitting in your favorite chair, this book can guide you through the heavens, revealing the gods, heroes, and monsters that people living long ago in the classical, Greco-Roman world believed resided there. Among them are Hercules, the greatest of all Greek heroes, and Perseus, slayer of snake-haired Medusa. Their number includes the winged horse Pegasus and Cetus, the sea monster sent to kill the princess Andromeda. These star myths have had such great staying power that the constellations named after them by the ancient Greeks are still known by the same names today.

The constellations and myths featured here are based on a list of forty-eight groups of stars contained in a book written almost two thousand years ago. The author of that book, entitled the *Almagest*, was a mathematician and astronomer known as Ptolemy, who wrote in Greek and lived in Alexandria, the most important city in Egypt at that time. Although written in the second century CE, the *Almagest* remained a highly valued textbook on astronomy until the Renaissance, and its catalogue of constellations is still used today, although Ptolemy's original list, based on observation with the naked eye, has now almost doubled.

I invite you to lie back, look up, and search the night skies for the serpents, dragons, bears, giants, strongmen, queens, kings, and princesses circling there.

INTROD

UCTION

I know well that I am mortal, a creature of one day.
But when my mind follows the wandering path of stars,
then my feet no longer rest on Earth, but standing next to
Zeus himself, I eat my fill of ambrosia, the divine food of gods.

—Ptolemy, *Almagest*, epigram

INTRODUCTION

STARS AND THEIR MYTHS

As long as humans have lived on Earth, objects in the sky have been a source of fascination to them. Looking at the night sky, they noticed that certain celestial bodies were brighter than others, that some appeared to remain in distinct groups or clusters, and that most of these changed in appearance, rising and setting at different times, as well as in different positions, throughout the year. They also noticed that these changes in the sky coincided with the progression of seasons, signaling when it was time to plant and when to harvest. Celestial bodies became critical tools for wayfinding on land and on the sea. Keeping track of the seasonal skies was a matter of survival.

What exactly these mysterious and powerful celestial objects were, early humans could not initially know. Nonetheless, people wanted and needed explanations for celestial phenomena, as for all phenomena in nature. It was this desire to understand nature that led to the creation of myths, stories firmly believed to be true and passed on orally from generation to generation, differing by cultural group. For the ancient Greeks, the Sun, Moon, and Earth were gods. Helios, the Sun god, drove his fiery chariot through the heavens, rising from his home beneath the seas in the morning and returning in the evening. The Moon goddess Selene, known to the Romans as Luna, drove her own bright chariot along the same course but did so at night. Gaia, the Earth goddess, remained stationary, of course, but became the source of every form of life.

Meanwhile, the stars beyond our solar system, which are less bright and for this reason somewhat "lesser" celestial lights, were not gods but rather heroes, heroines, animals, and monsters that the gods placed in the heavens after their deaths either as a reward or as eternal punishment. Looking up at star clusters in the night sky, people in the classical (Greco-Roman) world could see the great hunter Orion and the scorpion that killed him with its sting. They could see the centaur, half-man and half-horse, who raised the great Achilles. They could see poor Callisto, too. Callisto had once been a beautiful young woman but was transformed into a bear by the goddess Hera. Eventually, these myths were transmitted in writing, finding their way into texts of poetry and prose. Remarkably, they continue to live on even when their truth as definitive explanations for natural phenomena was challenged by thinkers who speculated, among other things, that the heavenly bodies were not gods or heroes but consisted instead of "matter"—air, water, or fire. These thinkers were philosophers, and their work would lay the foundation of natural science. Best

INTRODUCTION

known among them today are Pythagoras (who lived about 570–495 BCE), Plato (428/427 or 424/423–348/347 BCE), and Aristotle (384–322 BCE).

Who it was that first recorded the classical world's myths of the stars in writing remains a mystery. Not only this, but most classical authors whose work has survived do not tell all the stars' stories together, gathered in a collection, instead picking bits and pieces of well-known stories and using them as part of a larger narrative. In ancient Greece, writing was used in the Bronze Age (about 3000–1175 BCE) largely to keep track of precious food supplies and other resources but not to create literary works. Writing was a tool for recordkeeping. When Greece, under massive pressure from invaders from the north, subsequently plunged into a Dark Age, the art of writing was lost, only to re-emerge in Greece when people were able to become settled, living in relative peace. It was around 800 BCE that the Greeks adopted the alphabet of the Phoenicians, a wealthy, culturally advanced people whose territory extended from what is now the coast of Syria to Israel. The earliest piece of literature written using the new script was Homer's *Iliad*, a book-length epic poem centering on the exploits of the hero Achilles in the course of the Trojan War. The *Iliad*, the oldest extant literary work in the Western world (Europe), proved also to be its most influential, revered from the time of its composition and still a foundational work in popular culture today. Importantly, some details in the *Iliad*, which is conventionally dated to about 750 BCE, had their origins in the Bronze Age, when not one but a series of Trojan wars were waged. The adventures of Achilles and the other heroes who fought alongside him were first passed on orally, changing with each telling, until the time of Homer, whose own version was written down either by himself or by some other person's hand. That is, if there ever was in fact a Homer. Some scholars believe that Homer is perhaps not the name of a person at all but rather a name assigned to a tradition, the collective work of many poets.

Whatever the truth of Homer's existence may be, the Greeks themselves believed that it was he, and a somewhat later contemporary known as Hesiod, who gave them their gods, assigning the gods the names by which they would continue to be known, as well as the range of their powers. Zeus, for example, was the king of the gods and controlled the weather. He was the gatherer of clouds, the bringer of rain, and the source of both thunder and lightning. But he was also the most important civic god, guaranteeing that law and order prevailed, that people honored their promises, and so forth. Homer was no astronomer, but the *Iliad* contains a snapshot of how the universe in Homer's time was conceived. This appears in a detailed description of what was depicted on Achilles's shield. Homer describes Achilles's shield as picturing a round, flat Earth, circled by the great

INTRODUCTION

river Ocean, a massive river of fresh, not salty, water. Above the Earth, depicted on the domed boss at the shield's center, was the globe of the heavens, containing the Moon, Sun, and several constellations, namely, Orion and Ursa Major (*the Greater Bear*), as well as the Hyades and Pleiades, which are recognized today as two groups of stars in the constellation Taurus (*the Bull*). In mythology, the Hyades and Pleiades were daughters of Atlas, the Titan god whose job it was to hold up the heavens on his shoulders. Hesiod, for his part, described the origins of Gaia, goddess of the Earth, and the sky god Ouranos (Latin "Uranus") in his *Theogony*, a poem dedicated to the birth of all the gods. Hesiod is said to have written a poem about the stars, too, but that work, like so much written in antiquity, has not survived the passage of time.

There are as many variations of the classical star myths as there are authors who recorded them, and the myths appearing in this book are a synthesis of the best-known versions. Among the most important ancient sources for myths about the constellations is the Roman poet Ovid (lived 43 BCE–17/18 CE), active about 800 years after Homer's *Iliad*, was first committed to writing. Like all well-educated Romans, Ovid knew his Homer well, and he knew the work of other important Greek and Roman poets that lived before his own time as well. He drew freely but innovatively on the work of his predecessors. Ovid's best-known work is the *Metamorphoses*, like the *Iliad*, a book-length poem. Taking a page out of Hesiod's playbook, Ovid begins this work with an account of the origins of the universe, but the overarching theme of his poem is metamorphoses in which gods, humans, animals, and monsters change their shape—gods to animals or humans; humans to animals, plants, or stars; animals and monsters to stars and star-groups, too. More star myths, including alternate versions of those recorded by Ovid, can be found in the works of Hyginus, Ovid's contemporary, as well as in those of Apollodorus (writing in the second century CE) and Nonnus (writing in the sixth century CE), all of them collectors of myths. Valuable bits of information can be found in the works of other authors, too. Among these are the Greek historian Herodotus (fifth century BCE), the Greek geographer Strabo (another of Ovid's contemporaries), and the Roman travel writer Pausanias (second century CE).

While the star myths in this book combine details offered by a range of voices from classical antiquity, the book's organizational structure is based on a single ancient text, known today as the *Almagest*. It is neither a poem nor a collection of myths. It is not a book of history or an ancient travel guide. Instead, the *Almagest* is a scientific work that became the standard textbook on astronomy for well over a thousand years, from the time of its publication in the second century CE to the Renaissance. The *Almagest* contains a list of forty-eight constellations, all of them

INTRODUCTION

with the names of mythological characters by which they had long been known. Their stories are gathered in the pages that follow.

PTOLEMY AND HIS CONSTELLATIONS

The author of the *Almagest* is known today as Ptolemy, but his full proper name was Claudius Ptolemaius. Given how influential his book would become, it is surprising that so little is known about Ptolemy's life. In fact, we know almost nothing about him that is absolutely certain. The name Claudius indicates that he was Roman, while the name Ptolemaius points to a Greek-Egyptian ancestry. Clues detected in his surviving works suggest strongly that he lived and worked in the city of Alexandria in Egypt from approximately 100–174 CE.

Egypt had become part of the Roman Empire in the year 30 BCE, when the emperor Augustus defeated Cleopatra, Egypt's famously seductive last queen, and her lover, the Roman general Mark Antony. Founded by Alexander the Great in 332/331 BCE as the new capital of Egypt, Alexandria came to have enormous economic and cultural importance. It would become the "second city" of the Roman Empire, being second only to Rome itself. Alexandria was the main port of the eastern Mediterranean and central to an ever-increasing trade in luxury goods between India, Arabia, and Rome. A cultural melting pot, it was also a center of the arts and sciences, with a research institute called the Museum (Place of the Muses) that attracted a host of scholars over the centuries. By the time of Ptolemy, the city had what was likely still one of the best libraries in the world. It was an ideal place to engage in the kind of scientific research that Ptolemy so avidly pursued.

As a scholar, Ptolemy was interested in music theory, mathematics, geography, and astronomy, and he wrote a number of books on these topics, not all of which have survived. Of all his works, the *Almagest* is the most famous. The *Almagest* can be firmly dated to the reign of the Roman emperor Antoninus Pius (138–161 CE), the last of the so-called Five Good Emperors, whose rules were marked by prosperity and peace. Like much scholarly and scientific work in classical antiquity, the book was written in Greek, and its original Greek title, *Mathematikē Syntaxis* (Systematic mathematical treatise), reflected how greatly mathematics underpinned astronomical understanding. Significantly, it is the only surviving comprehensive ancient treatise on astronomy.

The *Almagest* is divided into thirteen chapters or books, as they were called in antiquity. It is a complete manual of astronomy intended for serious students of

INTRODUCTION

astronomy with a good working knowledge of geometry and familiarity with some basic astronomical terminology. Logically enough, it begins with the relationship of the heavens to Earth. In his words, "The heaven is spherical in shape and for this reason, heavenly objects move through the ether in a circular path.... It makes sense that the Earth as a whole is likewise spherical in shape ... and as for its position, it lies at the very center of the heaven" (*Almagest*, 1.10–20). Proof that the heavens are spherical, an enormous globe with Earth at its center, was the observation that "from their rising to their setting, the Sun, Moon and other stars are borne along in circular paths, always running parallel to each other, rising up from below as if from the depths of the Earth itself and ascending little by little, then proceeding along a similar arc, descending gradually until they appear to fall to Earth and vanish completely—only later, after remaining invisible for a while, to rise and set anew. Further, the duration of these cyclical motions, and also the locations of rising and setting, are, on the whole, fixed and the same" (*Almagest*, 1.10). Proceeding from these assumptions, Ptolemy then launched into a detailed, mathematically based discussion of the motions of the Sun, Moon, Mercury, Venus, Mars, Jupiter, Saturn, and the stars through the sky above us. The section on the stars includes a star catalogue, a list of 1,022 stars arranged into forty-eight constellations, with the position (location "on" the heavenly sphere measured by longitude and latitude) and magnitude (brightness measured on a scale from 1 to 6) of each. These forty-eight constellations were all given names, and these names were all taken from Greek mythology.

The influence of the *Almagest* on the history of astronomy cannot be overstated, even if not all of Ptolemy's astronomical observations were original to him. We know that he relied on advances made by earlier thinkers, including the Greek natural philosopher Hipparchus (active from about 150–127 BCE), from whose writings Ptolemy had learned a great deal. Hipparchus, for example, believed that the Earth's position had to be eccentric, not precisely at the center of the Sun's circular path around it, as this would account for the varying length of the seasons. Hipparchus also compiled a star catalogue, which Ptolemy greatly expanded. Other, even earlier predecessors whose work influenced Ptolemy included Anaximenes (lived about 585–528/4 BCE), who believed that all the stars were fixed in a crystal sphere that revolved around the Earth, and Eudoxus (lived about 400–347 BCE), who closely studied planetary movements and deduced that each planet was located in its own spinning sphere of air, each sphere being nested in the other like a Russian matryoshka doll. The starry sphere was the outermost. Around 500 BCE, Pythagoras proposed that the Earth itself is spherical, a notion fully accepted by Aristotle, seeing this as the best explanation for why some

INTRODUCTION

constellations, and not others, were visible from places at varying distances from the equator.

Though it was a groundbreaking text in its time, we also now know that Ptolemy's book contains errors. For example, the universe, or "heaven" as Ptolemy called it, is not a sphere (or a series of concentric spheres), and the positions of the stars that Ptolemy so carefully noted are not quite right. These errors, detected over millennia with enormous technological advances in the natural sciences, do nothing to diminish the importance of Ptolemy's book. Soon after its publication, it became the standard textbook of astronomy in the Roman world. As a reflection of the esteem in which it was held, the book began to be called "The Greatest Treatise" in place of "Systematic Mathematical Treatise," the title that Ptolemy had given it. In fact, the book would remain the dominant text on astronomy until the end of the sixteenth century, almost a millennium and a half after its initial publication. In about 800 CE, this "greatest" treatise was first translated into Arabic, and it is the Arabic title, *Almagest*, by which the book is still known today. Those who translated the book into Arabic transformed the Greek *hē megistē* ("the greatest") into the Arabic *al-majisti*, which in turn later became *almagesti* or *almagestum* in medieval Latin translations.

After the fall of the Roman Empire in 476 CE, the *Almagest* was lost to the western world until 1175, the high Middle Ages, when it was translated into Latin from Arabic by Gerard of Cremona. Born in northern Italy, Gerard traveled to Toledo in the Kingdom of Castile, where he learned Arabic and had access to a great many books in Arabic on astronomy, medicine, and other sciences. Knowing of the *Almagest*'s reputation as an authoritative text, he was keen to translate this work, and his translation became the most widely known in the western world until the Renaissance. The *Almagest*, meanwhile, remained the basis of astronomy in Europe until the publication of Polish astronomer Nicolaus Copernicus's *On the Revolutions of the Celestial Spheres* in the year 1543. Copernicus focused on what he thought were inconsistencies in Ptolemy's theory of an Earth-centric universe and formulated a new, revolutionary model of the universe that placed not Earth but the Sun at its center. In spite of this challenge, Ptolemy's influence was not broken until Danish astronomer Tycho Brahe (1546–1601) compiled the first star catalogue based entirely on independent observation rather than simply making the odd adjustment to Ptolemy's work. By insisting on meticulous firsthand observation, his measurement of the constellations' locations and the brightness of the stars that formed them was much more accurate. Still, it was not until the circulation of Brahe's assistant Johannes Kepler's observations on the elliptical

INTRODUCTION

motions of the planets around the Sun that the *Almagest* and Ptolemy's work on the movement of celestial bodies became obsolete as a textbook.

Nonetheless, even when many of Ptolemy's theories had been disproved, his catalogue of constellations remained a foundational text of observational astronomy. The International Astronomical Union (IAU) today recognizes and has catalogued eighty-eight known constellations. These are a combination of the forty-eight constellations recorded by Ptolemy, including their original Greek-mythological names, though translated into Latin, and forty additional star-groupings observed by later astronomers.

READING THE HEAVENS

In this book, Ptolemy's forty-eight constellations are arranged alphabetically, grouped by the region of the night sky in which they appear from our vantage point on Earth. These regions are called Quadrants and are divisions of the so-called Celestial Sphere. The Celestial Sphere is an imaginary hollow sphere with Earth at its center and the heavens "projected" onto its inner surface, a model very similar to how Ptolemy believed the heavens were configured. The Celestial Sphere's poles are aligned with Earth's north and south poles, and its equator is aligned with Earth's equator. The Celestial Sphere has northern and southern hemispheres, too, which are aligned with those of Earth. For purposes of mapping and reading the heavens, both Celestial Hemispheres are divided into four sections: Northern Quadrants 1, 2, 3, 4 and Southern Quadrants 1, 2, 3, 4. Imagine an orange cut in half and each half subsequently cut into four equal sections.

It is important to note that the Quadrants of the imaginary Celestial Sphere are different from what are called Galactic Quadrants. The latter can also be used to organize the constellations but are based on the Sun, not Earth, as their central point of reference. There is some overlap between Celestial and Galactic Quadrants, but not in the case of every constellation. Only Celestial Quadrants are referenced in the pages that follow. It is also the case that constellations themselves can be understood very differently by astronomers and more casual stargazers. The word "constellation" comes from the Latin word *constellatio*, which means "a grouping of stars," a combination of *con*, a variant of *cum* (meaning "with," "together") and *stella* ("star") together with the suffix *-tio* denoting the result of an action, in this case the gathering together of stars. Historically, constellations have been understood as recognizable patterns of bright stars. However, the IAU, in its

INTRODUCTION

effort to map the entirety of the heavens visible to us, has defined a constellation not as a distinctive grouping or pattern of bright stars but rather as a specific, measurable region of the sky containing such distinctive stellar patterns as well as other deep sky objects such as star clusters, nebulae, and galaxies in that region. For example, the IAU constellation Andromeda contains not only the pattern of bright stars that the Greeks imagined to be an image of the princess Andromeda in chains but also the Andromeda Galaxy, the Blue Snowball Nebula (NGC 7662), and an open star cluster (C28). This book focuses on constellations in the traditional, historical sense of recognizable patterns of bright stars visible to the naked eye. As was typical of historical star maps, lines linking these bright stars have been added in our illustrations as a tool to assist readers in the identification and visualization of the constellations' mythological figures. As an additional aid, the stars at the constellations' core are differentiated in accordance with their relative brightness viewed from Earth on a scale from 0 to 6, brightest to least bright, as follows:

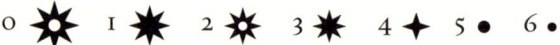

Because of Earth's own rotation, the Celestial Sphere appears to revolve around us carrying the constellations across the heavens overhead, rising in the east and setting in the west. Because the Earth not only rotates on an axis but also orbits the Sun, many constellations are visible only for part of the year, this period of visibility corresponding quite closely to the progression of seasons. Northern Celestial Quadrant 1 constellations are visible fall to winter, those of Northern Celestial Quadrant 2 are visible winter to spring, those of Northern Celestial Quadrant 3 are visible spring to summer, and those of Northern Celestial Quadrant 4 are visible summer to fall. Meanwhile, Southern Celestial Quadrant 1 constellations are visible spring to summer, Southern Celestial Quadrant 2 constellations are visible summer to fall, Southern Celestial Quadrant 3 constellations are visible fall to winter, and Southern Celestial Quadrant 4 constellations are visible winter to spring.

Additionally, constellations closer to the Celestial Equator are visible longer than those closer to the Celestial Poles. Northern Celestial Quadrant constellations are largely visible only from Earth's northern hemisphere, and, correspondingly, Southern Celestial Quadrant constellations tend to be visible largely from Earth's southern hemisphere. However, those constellations very close to or "on" the Celestial Equator are visible from both of Earth's hemispheres.

INTRODUCTION

The Sun, like the constellations, is imagined as being located on the Celestial Sphere's inner surface except that it is not "fixed," instead traveling in a yearlong circuit and passing through some of the constellations as it goes. The constellations through which the Sun appears to pass are the constellations of the Zodiac: Aries (*the Ram*), Taurus (*the Bull*), Gemini (*the Twins*), Cancer (*the Crab*), Leo (*the Lion*), Virgo (*the Virgin*), Libra (*the Scales*), Scorpius (*the Scorpion*), Sagittarius (*the Archer*), Capricornus (*the Sea-Goat*), Aquarius (*the Water-Bearer*), and Pisces (*the Fishes*).

FURTHER NOTES FOR THE READER

Most of the myths in this book are Greek, even those collected and retold (or reimagined) by Roman authors. It is also the case that, because of the enormous size and diversity of the Roman Empire, "Roman" authors were either native speakers of Greek or were fluent in Greek, which was considered to be more suited to intellectual and creative pursuits than Latin. For this reason, I have primarily used the Greek names of characters in my own retelling of the constellation myths, even where I have drawn on both Greek and Roman sources for a particular myth. There are a few exceptions, the most significant one being the hero Hercules. Hercules is the romanized (Latin) version of the hero's original Greek name, Heracles. Since most readers will know Hercules by his romanized name, that is the version that appears here. For those who are curious about the Roman equivalents of the Greek gods, or need a reminder of them, a list of them is included in this book's appendices.

While this book is a collection of what are primarily Greek myths, it is important to remember that the Greeks were by no means alone in mythologizing the stars. As mentioned earlier in this introduction, people from every cultural group have looked up into the night sky and observed the stars, recognizing distinctive groupings of them. Making and keeping track of such observations was a matter of survival, as the changing positions of these groupings signaled seasonal change and shifts in the weather, which had dramatic consequences for agriculture, travel over land or sea, and life more generally. Different peoples assigned these groupings different names and mythologies, all of them deeply interesting: the traditional star lore of New Zealand's Māori or that of the diverse peoples of Africa, India, Egypt, Mesopotamia (Babylonia), China, the Americas, and Scandinavia, to name just a few.

INTRODUCTION

Exactly what inspired the Greeks and others to see particular characters in the night skies is impossible to know. Why did Perseus holding Medusa's severed snake-haired head appear in the night sky and not the great Achilles, sword and shield in hand? Did the arrangement of stars really suggest a specific character, or were star myths mnemonic devices to help people remember them? Whatever the reason, it was often the case that a major figure was envisioned as being surrounded by other characters from the same myth. Every main character from the Perseus myth appears in the heavens: the princess Andromeda whom Perseus rescued, the sea monster (Cetus) from whom she was rescued, Andromeda's father, King Cepheus, and her mother, Cassiopeia. Hercules is accompanied by a number of the monsters that he slew or outwitted in the course of his famous Twelve Labors: the many-headed Hydra, Ladon the dragon (Draco), the Nemean Lion (Leo), and the giant crab (Cancer). He is accompanied by one of his weapons, an arrow poisoned by the Hydra's blood (Sagitta), as well as by the centaur (Centaurus) fatally wounded by that arrow's tip. Orion appears with his hunting dog (Canis Major) and with a rabbit (Lepus) that he was hunting. And so on. Clearly, the choice of one character could and did influence the choice of others from the same narrative, but not always.

An especially curious constellation is Capricornus, the Sea-Goat, envisioned as a goat with the tail of a fish. It is almost certainly the case that this hybrid creature was directly inspired by ancient Babylonian observations and myths of the stars and constellations. Babylonian star catalogues mention such a hybrid creature, a representation of the sea god Ea. Greek and Roman myths surrounding the sea-goat appear to have been fabricated specifically to explain the inclusion of a fishy goat among the constellations known by the Greeks and Romans. The Greeks and Romans did not exist in a cultural vacuum, instead being in constant contact with neighboring peoples. The exchange of goods and ideas in the broader Mediterranean was a robust one in antiquity, so it is unsurprising to detect outside influences or borrowings both in Greco-Roman advancements in astronomical knowledge and in classical star mythologies.

Finally, readers of this book may wonder at the violent nature of many of the myths, all of which in some way involve the gods. They may also wonder at the many instances of gods behaving badly—that is, in a way that we would today consider immoral, unethical, and criminal. The gods rewarded humans for good behavior and piety, but not always. Sometimes innocent characters were relentlessly persecuted for no fault of their own. Hercules is such a character. He was one of the many illegitimate children of Zeus, and because of this he was persecuted throughout his life by Zeus's jealous wife Hera. Hercules himself had done nothing wrong. Another example is the maiden Callisto, who steadfastly remained a virgin and was utterly devoted to the goddess Artemis. Zeus pursued and raped her, shamelessly taking on the appearance of Artemis to get close to her. Callisto was blameless, yet incurred the wrath of Artemis and of Hera as a consequence of Zeus's assault and was transformed into a bear. Zeus, on the other hand, shifted his largely unwelcome attention from one unsuspecting mortal woman to the next but without any consequence to him. What to make of this? The gods of Greco-Roman mythology were not perfect, ethical beings to be worshipped for their faultlessness. Instead, the gods were conceived of as not only being human in appearance but also as having human emotions and desires. Put differently, what distinguished gods from humans was primarily the fact that they were immortal, omnipotent, and (largely) omniscient. In his great epic poem the *Metamorphoses*, Ovid was clearly asking how humans could possibly revere deities who behaved in such a way. And yet they did, perhaps precisely because they were reflections of humans at their best and at their worst.

PTOLEM
CONSTEL

Y'S

LATIONS

NORT
CELE
QUAD

HERN
STIAL
RANT
I

NORTHERN CELESTIAL QUADRANT I
Andromeda

NORTHERN CELESTIAL QUADRANT I

ANDROMEDA

Andromeda

As soon as Perseus caught sight of her there, her arms shackled to
a bristling crag, he was instantly consumed with passion's flames,
struck dumb, captivated by her beauty. Indeed, had it not been
that her hair stirred ever so slightly in the breeze, and her cheeks
glowed warm and wet with tears, he would have
thought her a breathtaking marble statue.

Ovid, *Metamorphoses* 4.672–676.

MAIN CHARACTERS:
Andromeda, princess of Ethiopia
Athena, Greek goddess of wisdom and defensive war
Cassiopeia, queen of Ethiopia and mother of Andromeda
Cepheus, king of Ethiopia and father of Andromeda
Cetus, a sea monster
Medusa, one of the three Gorgon sisters
Nereids, nymphs of the sea
Perseus, Andromeda's rescuer and Medusa's slayer
Poseidon, Greek god of the sea
Zeus, king of the Greek gods

Andromeda was a princess of Ethiopia and famed for her beauty. Cepheus, Ethiopia's king, was her father, and Cassiopeia, the queen, her mother. Cassiopeia had good reason to be proud of her daughter's looks, but her natural, maternal pride turned tragically to hubris. Her daughter, Cassiopeia exclaimed publicly, was more lovely even than the Nereid nymphs, spirits of the seas. At this boast, the Nereids took affront, and their watery neighbor, the sea god Poseidon, came to their defense.

As punishment for the queen's misdeed, all of Ethiopia was made to feel Poseidon's wrath. The god sent an enormous tidal wave to flood the kingdom's lands and a sea monster, the fearsome Cetus, to ravage the people and their livestock. A distraught King Cepheus now traveled into the Libyan desert to consult the famed oracle of Ammon, where he learned that the only remedy for

CLASSICAL MYTHOLOGY OF THE CONSTELLATIONS

this calamity was the sacrifice of his daughter, Andromeda. So it came to pass that Andromeda was shackled to the shore's rocky crags as a lovely morsel for Poseidon's monster.

As the princess awaited her fate, it happened that the hero Perseus, Medusa's severed head in hand, passed just overhead, borne through the air on winged sandals. Though hastening to the island of Seriphos to save his mother from being forced to marry that island's evil king, Perseus came to a halt midair, transfixed by the sight of Andromeda. He was instantly in love. A bargain was struck with Cepheus: Perseus would slay the monster and, as compensation, receive Andromeda as bride. The deed was quickly accomplished, but now Perseus faced an unexpected obstacle to claiming his beloved. Andromeda, it was revealed, had been promised to Cepheus's brother, Phineus, her own uncle. Andromeda's clear preference was Perseus. She loved her savior with all her heart. But Phineus, angry at being slighted, would not give her up. In his view, a promise was a promise. Nor, indeed, would Perseus step aside. A peaceful solution could not be found, and Perseus had the perfect weapon to use against his formidable new enemy. Bursting in upon Phineus and his friends when they were gathered for a feast, Perseus transformed them all into stone by holding up Medusa's gruesome, snake-haired head.

Andromeda went with Perseus to Seriphos and, thereafter, to the cities of Argos, Larissa, and Tiryns in mainland Greece. Perseus would ultimately become king of the powerful kingdom of Tiryns, and Andromeda the kingdom's queen. A model wife and mother, she bore Perseus a daughter and six sons. At her death, the goddess Athena, who had a special bond with Perseus, placed her among the stars in recognition of her virtue. There, in the heavens, she was joined by Perseus, her parents, and the sea monster Cetus, all of them given eternal life as clusters of stars.

Andromeda, one of the best-known constellations, depicts the princess in chains, just as she was when the hero Perseus discovered her. Largely confined to Northern Quadrant 1, the starry priness Andromeda reaches into Northern Quadrant 4 with her right arm. This constellation is especially notable for containing the galaxy closest to Earth, the so-called Andromeda Galaxy (M31). Located at a distance well over two million light-years from Earth, the Andromeda Galaxy is the most distant celestial object visible without a telescope.

Best visibility: October to November (90°N–37°S)

See also the constellations Cassiopeia, Cepheus, Cetus, Pegasus, and Perseus.

NORTHERN CELESTIAL QUADRANT I

ARIES

The Ram

> Nephele, learning that her children would be killed, seized both young Phrixus and her daughter Helle, and gave them a golden-fleeced ram that she had received from Hermes, by which they were carried through the sky, over land and sea. But as they reached the sea that lies between Sigeum and the Chersonese, Helle slipped, falling to her death into the waves.
>
> Apollodorus, *Bibliotheca* 1.9.1

MAIN CHARACTERS:
- Aeetes, the king of Colchis
- Ares, god of war
- Athamas, a king of Orchomenos and father of Phrixus and Helle
- Helle, Phrixus's younger sister
- Hermes, the messenger god and protector of travelers
- Jason, the Greek hero sent to retrieve the golden fleece
- Nephele, first wife of Athamas
- Phrixus, son of Athamas by Nephele
- Zeus, king of the gods

The constellation Aries (*the Ram*) takes its name from the ram whose fleece was sought by the hero Jason and the Argonauts, the brave sailors on the ship *Argo*. Needless to say, this was no ordinary ram. It possessed the ability to fly and had a fleece of precious gold. This ram was given by the god Hermes to Nephele, queen of Orchomenos in Boeotia, in central Greece, and what follows is the ram's story.

Nephele was the first wife of King Athamas, who proved to be among the unluckiest of men. Poor judgment led him to take another wife while Nephele was still alive. This new wife, Ino, bore him two sons, but Ino feared that Nephele's son, Phrixus, would become Athamas's heir, leaving her sons with nothing.

Not daring to confront Athamas, Ino hatched a plot by which to eliminate Phrixus. She ordered the kingdom's store of seed grain to be scorched, which ensured that when it was sown in spring, it subsequently failed to sprout. Famine would be the certain outcome. In alarm, King Athamas sent an envoy to Delphi

NORTHERN CELESTIAL QUADRANT I

Aries

to ask the oracle of Apollo how famine could possibly be avoided. The envoy, however, did not report the god's true words. On his return from Delphi, he was intercepted by Ino and bribed to utter words composed by Ino herself: human sacrifice was the only remedy, and the sacrifice needed to be Phrixus, Nephele's son. Athamas was understandably reluctant, but there seemed to be no alternative, so he readied himself to make the required sacrifice. Fortunately, Nephele was able to snatch Phrixus from a certain fate at his father's hands. Holding fast to her children, Phrixus and his sister, Helle, she placed them on the back of a golden-fleeced, flying ram that miraculously appeared, a gift to Nephele from the god Hermes. This ram would carry them to safety. However, as the ram flew over the strait separating Europe from Asia, Helle lost her grip and fell to her death in the waters below. The waters in which she drowned were named Hellespont ("Helle's Sea") after her. The Hellespont is now called the Dardanelles.

Phrixus, meanwhile, was carried safely to the kingdom of Colchis, which lay on the shores of the Black Sea. Aeetes was that land's king. When he arrived, a grateful Phrixus sacrificed the ram to Zeus. To commemorate its service, the ram was lifted to the heavens by the gods in effigy, appearing now as a constellation, and its fleece was hung upon an oak tree in a grove sacred to the war god Ares. There the fleece was guarded by a sleepless dragon, remaining secure until the hero Jason came on the ship *Argo* (Argo Navis) to take it back to Greece. As for Phrixus, he married one of King Aeetes's daughters, gave her many children, and lived to a ripe old age.

Best visibility: November to December (90°N–58°S). Aries is one of the twelve constellations of the Zodiac, with the Sun passing through it from mid-April to mid-May.

See also the constellation Argo Navis.

CLASSICAL MYTHOLOGY OF THE CONSTELLATIONS

AURIGA

The Charioteer

Over there are men shouting a warning for the Pisan king—
maybe you can hear them—and the setting is Arcadia, a region
of the Peloponnese. The chariot, drawn by a team of four horses,
soon lies shattered through a trick of Myrtilus.

Philostratus, *Imagines* 1.17

MAIN CHARACTERS:

Agamemnon, leader of the Greek forces in the Trojan War and descendant of Pelops
Amalthea, the she-goat said to have nursed an infant Zeus
Athena, the patron goddess of Athens
Cecrops, a legendary king of Athens
Clymene, the mother of Phaethon
Demeter, the Greek goddess of grain and the harvest
Erichthonius, a snake-bodied child of Athena and Hephaestus
Helios, the Sun god and father of Phaethon
Hephaestus, god of volcanic fire and the forge
Hermes, the messenger god and father of Myrtilus
Hippodamia, the lovely princess of Pisa
Hippolyta, an Amazon queen and mother of Hippolytus
Hippolytus, the son of Hippolyta and the hero Theseus
Myrtilus, the charioteer of Pisa's evil king Oenomaus
Oceanus, a Titan god and the freshwater river believed to circle the Earth
Oenomaus, the king of Pisa defeated by Pelops in a chariot race
Orestes, the son of Agamemnon
Pelops, the son of Tantalus and later king of Pisa
Persephone, the daughter of Demeter abducted by Hades
Phaedra, the wife of Theseus and stepmother of Hippolytus
Phaethon, the son of Helios who came to grief driving his father's chariot
Poseidon, Greek god of the sea
Tantalus, a man once loved by the gods but eternally tortured in the afterlife

Theseus, a legendary king of Athens and father of Hippolytus
Thyestes, a tragedy-stricken descendant of Pelops

A number of charioteers featuring in classical mythology became associated with the large constellation Auriga (*the Charioteer*). The most notable of these were Myrtilus, the treacherous charioteer of Pisa's king Oenomaus; Hippolytus, a grandson of the god Poseidon; Phaethon, who drove the Sun god's chariot; and Erichthonius, the legendary inventor of the four-horse chariot.

Myrtilus, a son of the god Hermes, played a critical role in the story of the hero Pelops. Pelops was the son of the infamous sinner Tantalus, once a favorite of the gods but a man who abused his privileged status. Tantalus invited the gods to dine with him and, in order to test whether they were actually omniscient, cut up his own son Pelops and served him to the gods in a stew. The gods sensed the foul and sacrilegious nature of the meal before them, except Demeter, who was too distracted by grief over the recent loss of her daughter Persephone to notice anything. She had already taken a bite, which proved to be part of Pelops's shoulder. In any event, a horrified Zeus pieced Pelops back together, inserting a shoulder of ivory for the missing body part, and restored him to life. For his sins, Tantalus was doomed to eternal punishment in the Underworld: he stood in a pool whose waters retreated whenever he stooped to drink; a branch heavy with fruit hung just overhead, but this would retract whenever he reached up in hunger; and an enormous stone balanced over him, always threatening to fall.

As for young Pelops, when he was old enough to marry, he became enamored with Hippodamia, the lovely daughter of Oenomaus, the king of Pisa. Oenomaus, as it happened, loved his daughter, too, so much so that he was unwilling to see her wed. In order to prevent her marriage to anyone without betraying his own unnatural desire for her, he challenged each of her suitors to a chariot race, a race that he would be destined always to win. With his charioteer Myrtilus at the reins, Oenomaus, riding alongside, would shoot his adversaries with his bow and arrow if they should come too near. When Pelops arrived at Pisa, a dozen suitors had been killed, their heads impaled on posts in front of Pisa's palace as a deterrent to any future suitor. Pelops remained undeterred, however. He bribed Myrtilus with an offer of half the kingdom and a night with Hippodamia for his help, and Myrtilus enthusiastically agreed. Then, as the king's chariot was being readied, Myrtilus removed the linchpins that secured the vehicle's wheels to its axles. Meanwhile, Pelops leaped into his chariot with Hippodamia at his side. In alarm, Oenomaus himself grabbed the reins of his own chariot and went in pursuit. As soon as his chariot reached speed, the wheels came loose, and the king was thrown, fatally wounded. Pelops now was king of Pisa, and he had won his precious bride. But when Myrtilus reminded him of the payment due, Pelops, having no intention of keeping his own promise, pushed him from the nearby cliffs. Before falling into the sea where he drowned, Myrtilus uttered a curse on Pelops and all of his descendants. These descendants included Agamemnon, victorious commander of the assembled Greeks who fought (and won) the Trojan War, but who was later killed gruesomely by his wife. His descendants also included Agamemnon's adulterous brother Thyestes and Agamemnon's son Orestes, who became a matricide. As for Myrtilus, Hermes placed his image in the heavens even though his deeds had been far from honorable.

Hippolytus, another of the charioteers linked with the constellation Auriga, also met an untimely end. Hippolytus was the illegitimate son of Theseus, a legendary king of Athens, and of the Amazon queen Hippolyta. Hippolytus's misfortunes began when Phaedra, Theseus's wife, fell hopelessly in love with him. He rebuffed her advances and, filled with shame, she committed suicide, leaving behind a note accusing Hippolytus—falsely—of having pursued her. In a rage, Theseus cursed his son, praying for his death. In answer to this prayer, the god Poseidon sent a bull from the sea just when Hippolytus was driving by in his chariot. Hippolytus's spooked horses bolted, throwing him from the chariot and dragging him to his death tangled in the reins. Hippolytus would then live again, transported to the skies by the gods as a constellation.

Misfortune and a tragic end were part of Phaethon's story, too. He was the child of the Sun god Helios and Clymene, known variously as a nymph and as a daughter of the Titan god Oceanus. After giving birth to Phaethon, Clymene married an Egyptian king who raised Phaethon as his own. Young Phaethon, however, learned that the king was not his father and insisted on discovering his true parentage. His father was Helios, he was told. So Phaethon made his way to Helios's palace, where he was welcomed by the god and offered any favor as proof that Helios was actually his father. To take the reins of Helios's blazing chariot was Phaethon's request. And, though worried for the safety of his son, Helios could not refuse. No god would break a promise. Phaethon began his fateful journey through the heavens well enough, but he lacked the experience and strength to control the chariot's team of horses. They galloped upward in the sky, blazing a milky scar in the heavens (the Milky Way) and causing the constellations to draw back in fear. Then they veered toward the surface of the Earth, scorching lands that then became deserts, together with their inhabitants, darkening their skin. With the heavens and Earth under threat, Zeus, reluctant though he was to harm

NORTHERN CELESTIAL QUADRANT I
Auriga

the youth, hurled a lightning bolt at Phaethon. His lifeless body fell from the sky into the river Eridanus. Both Phaethon and the river were commemorated by Helios with constellations in their image, the charioteer appearing in the skies driving his father's chariot.

Finally, there is Erichthonius, who is better known for his strange birth and appearance than for his deeds as king of Athens, however notable. His parents were the goddess Athena and Hephaestus, god of volcanic fire and of the forge, but he was conceived in a most unusual way. Hephaestus pursued Athena, but the goddess had no interest in him. When, in desperation, he resorted to rape, she pushed him forcefully away. In the struggle, a drop of semen had fallen into her lap, and this she brushed onto the earth. In due course, a child with the body of a snake emerged from the place where the god's seed had fallen. This was Erichthonius. Athena placed her newborn son in a chest and gave him to the daughters of Cecrops, king of Athens, to keep safe. Under no circumstances should they open the chest, Athena said. However, curiosity got the better of them, and they peeked inside. What they saw was terrifying, so much so that they leaped to their deaths from the walls of the Acropolis out of fear. The snake-bodied Erichthonius eventually succeeded to the kingship of Athens and was said to have instituted a great festival, the Panathenaia, in his mother's honor. According to some sources, the gods saw to it that Erichthonius, a noble king and inventor of the four-horse chariot, was raised to the heavens upon his death.

Curiously, Auriga's charioteer is depicted in the heavens without his chariot but holding up a whip in his right hand. More curiously still, there is a she-goat and her two small kids on his left arm. In mythology, the goat is identified as Amalthea, a goat who is said to have nursed Zeus when he was a baby.

Best visibility: December to February (90°N–34°S)

See also the constellation Eridanus.

NORTHERN CELESTIAL QUADRANT I

CASSIOPEIA

Cassiopeia

Because Cassiopeia, the wife of Cepheus, competed with the Nereids in beauty, boasting that she was lovelier than them all—this was the reason why the Nereids were angry, and Poseidon, sharing their resentment, sent a flood and a monster to invade the land.

———

Apollodorus, *Bibliotheca* 2.4.3

MAIN CHARACTERS:

Ammon, an Egyptian god identified by the Greeks with Zeus
Andromeda, an Ethiopian princess rescued by Perseus from the sea monster Cetus
Athena, the goddess of wisdom and defensive war
Cassiopeia, the queen of Ethiopia and mother of Andromeda
Cepheus, the king of Ethiopia and father of Andromeda
Cetus, the sea monster sent by Poseidon to ravage Ethiopia's coast
Medusa, the snake-haired Gorgon beheaded by Perseus
Poseidon, god of the sea

Cassiopeia was the wife of Cepheus, a king of Ethiopia. It was no noble deed but rather an indiscretion that landed her in the heavens among the stars. Cassiopeia was very beautiful, and this went to her head. Out of pride, and in a massive lapse of judgment, she compared herself to the nymphs of the seas, the Nereids, in beauty. Worse still, she claimed to be more beautiful than they. The Nereids, of course, were angered at this prideful boast. Sharing their outrage, their neighbor Poseidon unleashed a twofold punishment upon Cepheus's kingdom: a flood and a sea monster that set about ravaging the Ethiopian coast.

King Cepheus then traveled to the oracle of Ammon in Libya to inquire what remedy there might be for this disaster. What he learned made him desperately sad, but there seemed no other way: he was to sacrifice his own daughter, Andromeda, to the sea monster Cetus, chaining her to a towering face of rock along his kingdom's shore. However, by a stroke of good fortune, the hero Perseus came flying overhead on his way home after beheading the Gorgon Medusa.

NORTHERN CELESTIAL QUADRANT I
Cassiopeia

NORTHERN CELESTIAL QUADRANT I

Perseus happened to look down and spotted Andromeda, with whom he fell immediately in love. He resolved to save her and, at the same time, win her as his bride. So, he approached Cepheus and Cassiopeia, offering his help in exchange for their daughter's hand. They readily agreed, and Perseus did successfully rescue Andromeda, though he first had to fight Andromeda's uncle, to whom she had previously been promised, before taking her away with him.

At the end of their lives, the goddess Athena immortalized Andromeda and Perseus as constellations. Andromeda's parents and Cetus, the sea monster, became constellations, too. This was Poseidon's doing. In the case of Cassiopeia, however, her punishment continued, even in the heavens, for she was placed in a chair, tipped over, so that she would have to travel through the sky on her back with her feet in the air.

In a slightly different version of this tale, Cassiopeia claimed not that she herself but rather her daughter Andromeda was more beautiful than the Nereids, causing them and Poseidon to seek vengeance. The outcome was the same. A sea monster was sent to plague the land of Ethiopia, and the oracle of Ammon revealed that Cepheus would need to sacrifice Andromeda to rid the land of this monster. Perseus rescued and won the hand of Andromeda, and the whole cast of characters became constellations.

Best visibility: October to December (90°N–12°S)

See also the constellations Andromeda, Cepheus, Cetus, and Perseus.

ORION

Orion

Orion, who surpassed all the heroes ever known in size and bodily strength, loved hunting with a passion and was a builder of great works by virtue of his great might and his drive for glory.

Diodorus Siculus, *Library of History* 4.85.1

MAIN CHARACTERS:
  Artemis, goddess of the hunt and protectress of wild animals
  Cedalion, the young man who guided a blind Orion
  Euryale, a princess known by some as Orion's mother

CLASSICAL MYTHOLOGY OF THE CONSTELLATIONS

Gaia, the Earth goddess and mother of Orion
Hera, the queen of the gods
Hermes, the messenger god and one of the three fathers of Orion
Hyrieus, Orion's foster father
Oenopion, the king of Chios and Side's father
Orion, one of mythology's greatest hunters and a man of enormous size
Poseidon, god of the sea and one of the three fathers of Orion
Side, Orion's boastful and vain first wife
Zeus, king of the gods and one of the three fathers of Orion

The hero Orion, after whom this constellation is named, was one of classical mythology's greatest hunters and a man of giant proportions. It was said that he was taller than the sea was deep. He was also a curiously tragic, flawed figure. By some accounts, he was the son of a Cretan princess, Euryale, and the god Poseidon. Alternatively, he was known as the foster son of Hyrieus, a king of Boeotia in central Greece (or of Thrace, a region lying between the Balkan Mountains, the Aegean, and the Black Sea).

 King Hyrieus had been childless and very unhappy about not having an heir. It so happened that the gods Zeus, Poseidon, and Hermes were journeying among mortals and stopped at Hyrieus's palace. There they were entertained hospitably, and the grateful gods resolved to reward him. They would grant him any wish. Hyrieus's wish, of course, was for a son. So, the gods asked an unsuspecting Hyrieus for the hide of a recently sacrificed bull and then proceeded to urinate in it and bury it. After nine months, much to Hyrieus's surprise and delight, an enormous baby Orion emerged. Orion regarded Hyrieus as his father. His mother, technically, was Gaia, the deified Earth, in whose soil he had been buried and incubated.

 When Orion reached maturity, he married a young woman by the name of Side, but this union was short-lived because Side boasted that she was more beautiful than the goddess Hera, who punished her with death. Orion then made his way to the island of Chios, where he became enamored of King Oenopion's daughter. He was so smitten with her that he could not restrain himself and made improper advances on her. Her father called this rape. In his fury, Oenopion blinded Orion and cast him out alone onto the beach.

 From his divine "father" Poseidon, Orion had received the gift of being able to walk on water, so he made his way from Chios to the island of Lemnos, where he was given (or abducted) a young man to serve as his guide. This man, named

NORTHERN CELESTIAL QUADRANT I
Orion

Cedalion, rode on Orion's shoulders. The two journeyed to the farthest East, to the abode of the Sun god Helios, who cured Orion of his blindness.

After a number of adventures, among them an unsuccessful attempt at vengeance upon Oenopion, Orion arrived on the island of Crete, where he enjoyed the company of the goddess Artemis. Together, Orion and the goddess would roam the countryside and hunt, but this relationship, too, was ill-fated. Orion became so skilled a hunter that, in a moment of supreme overconfidence, he boasted of his ability to kill every animal on Earth. The prospect of such an event, or even the thought of it, was too much for the gods to bear, including Artemis, protectress of wild animals. In antiquity, hunting was never pure sport but, instead, an activity performed with reverence and respect for one's prey. Prayers and thanks were offered for every victim. So, alarmed at Orion's boast, the Earth goddess Gaia sent a giant scorpion to kill Orion with its sting before he could do the harm he had predicted.

Admiring the scorpion for its role in preserving the animal kingdom, Zeus placed it among the stars as a warning to humanity. The scorpion is the constellation Scorpius, and its supersized claws were a separate constellation (called Libra, *the Scales*, because many viewed it not as claws but as scales). A conflicted Artemis felt sorry now for Orion and asked Zeus to place Orion in the heavens, too, as a memorial to his bravery. And so it came to pass that both the scorpion and Orion appeared in the heavens as constellations, the one rising as the other sets. Orion's faithful hunting dog and a hare that Orion presumably pursues complete the starry tableau. Orion's favorite dog is generally viewed as the constellation Canis Major (*the Greater Dog*), while Canis Minor (*the Lesser Dog*) is another, the smaller of Orion's dogs. The hare, meanwhile, is the constellation Lepus (*the Hare*).

Orion, who is visible from everywhere on Earth, strides vigorously through the night sky, with a club (or sword) in his raised right hand and an animal skin draped over his left. Three very bright stars signify his belt. Orion looms so large in the sky that he spans the Celestial Equator, the stars of his belt and his lower body extending into Southern Quadrant 1.

Best visibility: December to January (79°N–67°S)

See also the constellations Canis Major, Canis Minor, Lepus, Libra, and Scorpius.

NORTHERN CELESTIAL QUADRANT I

PERSEUS

Perseus

He dove into that dangerous cave and keeping close against the walls of stone, he reaped a hissing harvest, first fruits of tangled hair, slicing the Gorgon's writhing throat and staining his sickle red.

―――――

Nonnus, *Dionysiaca* 31.17–20

MAIN CHARACTERS:
- Acrisius, king of the Greek city of Argos
- Andromeda, princess of Ethiopia
- Apollo, the Greek god of prophecy and healing
- Athena, Greek goddess of wisdom and defensive war
- Cassiopeia, Queen of Ethiopia and mother of Andromeda
- Cepheus, King of Ethiopia and father of Andromeda
- Cetus, a sea monster
- Chrysaor, a child of Medusa and brother of Pegasus
- Danaë, princess of Argos and mother of Perseus
- Dictys, a fisherman on the island of Seriphos and brother of Polydectes
- Erinyes, spirits of vengeance
- Graiae, the Gray Ones, three monster-sisters
- Hermes, the Greek messenger god
- Hippodamia, a princess of Pisa
- Medusa, one of the three Gorgon sisters
- Pegasus, a winged horse and child of Medusa
- Perses, a son of Perseus and Andromeda
- Perseus, Medusa's slayer and rescuer of Andromeda
- Polydectes, villainous king of the island of Seriphos and brother of Dictys
- Zeus, king of the Greek gods

The constellation Perseus represents the Greek hero who is best known for having beheaded the fearsome snake-haired Medusa. Perseus was the son of the princess Danaë and Zeus, king of the Greek gods. With these parents, his life should reasonably have been one of ease and privilege. Instead, it was filled with harrowing adventures from the very moment of his birth. The cause for these adventures, however, were events taking place well before Perseus was born.

Perseus's grandfather, King Acrisius of Argos, desperately wanted a son and heir, but his wife only bore him daughters. Wanting to know if he would ever have a son, he traveled to Delphi in northern Greece to consult the famous oracle of the god Apollo. What he learned there alarmed him thoroughly. The god's priestess revealed that no son of his own was in his future. The priestess went on to say that his daughter Danaë, instead, would have a son, but that this child was destined, one day, to kill him. In light of this news, ordering Danaë to be locked away seemed the best plan. Killing her was not an option because, in the Greek world, punishment for murder was death. Murder, especially killing family members, was sure to rouse the Erinyes—bloodthirsty, avenging spirits who would chase perpetrators to an early grave.

This is how it came to pass that the place of Perseus's birth and conception was an impenetrable chamber, secure enough to keep out all who might try to breach its walls—all, that is, apart from the gods. Zeus, who regularly changed his shape to seduce and sleep with mortal women, transformed himself into a shower of gold that could and did reach Danaë through the smallest of the chamber's openings. In the course of this encounter, Perseus was conceived.

When, much to his horror, Acrisius learned that his daughter was pregnant, he was faced with a dilemma. Again, directly causing Danaë's death would spell his own. So, when the child, a son, was born, Acrisius ordered Danaë and the newborn Perseus to be set adrift in a sturdy chest of bronze. The expectation was that they would die a natural death at sea. But the gods were watching over them, and the chest washed ashore on the island of Seriphos, where it was discovered by a fisherman called Dictys.

The kindhearted Dictys brought Danaë and Perseus to the safety of his house. It was there that they lived in peace until, one fateful day, Danaë caught the eye of Polydectes, Dictys's brother and the island's evil king. An infatuated Polydectes became determined to marry Danaë but considered young Perseus, who was fiercely protective of his mother, a potential threat.

Once again murder as a means of elimination was not an option, so Polydectes devised a ruse and a challenge that Perseus was unlikely to survive. Disguising his true purpose, Polydectes spread a lie, making it known that he was marrying Hippodamia, princess of Pisa. Those invited to the wedding were told that horses were what Polydectes wanted as a wedding gift. Horses, the most highly valued animals in ancient Greece, were the ultimate symbols of great wealth and power. Polydectes knew that Perseus, having been raised as a humble fisherman, would

NORTHERN CELESTIAL QUADRANT I

Perseus

never be able to offer such a costly gift, so he suggested an alternative. This was the head of Medusa, a prize he felt confident would be impossible for Perseus to obtain. Even if he managed to find Medusa, beheading her was unsurvivable.

Medusa was one of three Gorgon sisters who lived in a dank, dark cave at the very edges of the world. They were the stuff of nightmares, having fangs, claws of bronze, and hair of writhing snakes. Two of the sisters were immortal, but Medusa was not. Still, she was far from easy prey. Although she could be killed, anyone who so much as glanced at her turned instantly to stone. Perseus would need help, which the goddess Athena and the god Hermes freely gave. With their guidance, Perseus sought out the strange Graiae ("Gray Ones"), three grizzled old sisters who shared a single tooth and a single eye that they passed between them. They, Athena told him, would be able to direct him to the nymphs of the Far North who, in turn, would give him the equipment required to complete his formidable task. However, since the Graiae were the Gorgons' sisters, they would have to be tricked into exposing the Gorgons to danger and revealing what they knew. Perseus traveled to the Graiae and, following Hermes's instructions, he grabbed the Graiae's eye, in this way forcing the now-blinded sisters to reveal the way to the nymphs in exchange for their sight.

When, at last, he came to the nymphs' abode, these kind deities gave him a satchel in which to carry Medusa's head, winged sandals with which to make a speedy escape, and a special cap that made its wearer invisible. From Hermes he received a sword of the hardest adamant and from Athena a highly polished shield. Suitably equipped for his impossible task, Perseus approached the Gorgons' lair. As silently as possible, he crept inside, pressed close to the cave's icy walls. Although his cap made him invisible, he was naturally still afraid. Deep inside their cave, Perseus located the sisters huddled together and deep in sleep. This he did by using his shield as a mirror, seeing their reflections but never having to look at them directly. Moving swiftly and stealthily, he rushed at Medusa and lopped off her ghastly head. From the severed neck leaped the winged horse Pegasus, born fully grown, as well as a giant known as Chrysaor. Medusa, who at one time had been a beautiful maiden, was raped by Poseidon. The result of this violent encounter was Medusa's monster-children, their strange and awful conception resulting in an even stranger birth. It took their mother's beheading for them to be born at last.

Sword and Gorgon head in hand, Perseus then made for the cave's opening. Once outside, Medusa's head stashed inside his satchel, he fled, with Medusa's sisters, now wakened from their slumber, in pursuit. Borne aloft on his winged sandals, Perseus easily outpaced the Gorgons and sped through the heavens back to Seriphos. But, on his way, he caught sight of a lovely maiden, the princess Andromeda, chained to a rocky cliff as an offering to the dreaded sea monster Cetus. Andromeda's father and Ethiopia's king, Cepheus, had reluctantly given the order to sacrifice his daughter in order to save his kingdom from a great flood and the sea monster who was ravaging the city. In exchange for Andromeda's hand in marriage, Perseus slew the monster. The Ethiopians rejoiced. Their princess and their lands were saved.

Nonetheless, obstacles to Perseus and Andromeda's happiness remained. During the festivities that ensued, it was revealed that Andromeda had been promised in marriage to King Cepheus's brother. Perseus, it turned out, could not have Andromeda without a fight, a fight that he won easily by holding up Medusa's head when surrounded by his adversaries, turning them to stone.

With Medusa's head safely back in his satchel, Perseus now returned to Seriphos accompanied by his new bride. There he learned that the villainous King Polydectes had continued to pursue his mother, and that the wedding he had earlier announced was all a lie. In fear of Polydectes, his mother and Dictys had been forced to go into hiding. An angry Perseus set off for the palace, where Polydectes and his friends happened to be gathered for a feast. All eyes turned to Perseus in surprise as he entered the banquet hall. It was as if a ghost had appeared before them. Surely Perseus was dead! As they all looked at him in astonishment, Perseus again held up Medusa's head, turning the assembled company to stone.

Not wishing to assume the kingship of Seriphos, Perseus declared the fisherman Dictys king and, with Andromeda, he headed for Argos, land of his birth, to find his grandfather King Acrisius. Remarkably, Perseus bore Acrisius no ill will for casting him and his mother adrift to meet an uncertain fate. It was also high time now, he felt, to return the satchel, sandals, cap, and shield. He gave the head of Medusa to the warrior goddess Athena, who fixed it proudly to her breastplate.

Meanwhile, Acrisius had heard of Perseus's heroic deeds and, fearing that his grandson would come for him, fled the kingdom of Argos for Larissa in northern Greece. Even now he hoped to escape the god Apollo's prophecy. Discovering that Acrisius had gone, Perseus followed him, and it was at Larissa that the prophecy Acrisius so feared was fulfilled. It so happened that while Acrisius and Perseus were in Larissa, the father of that city's king passed away. As was customary in the Greek world, sporting events were held in honor of the deceased, and illustrious Perseus participated. While competing in the pentathlon, he hurled his discus, only—quite by accident—to strike Acrisius dead. Filled not with joy but with deepest shame over causing his grandfather's death, Perseus refused the kingship of Argos, instead making a trade with his cousin, the king of Tiryns, a city neighboring Argos.

As regents of Tiryns, Perseus and Andromeda became parents to a daughter and six sons, among them Perses, who would become the founding ancestor of the Persians. Perseus himself was remembered as one of Greece's greatest heroes, and upon their deaths, Athena placed Perseus and his wife Andromeda among the stars, where they would be immortalized as constellations neighboring Cepheus, Cassiopeia, and the sea monster Cetus. Even Pegasus, Medusa's miraculous child, would find a place in the heavens as a constellation. As for Perseus, he appears in the sky brandishing his sword and clutching Medusa's head.

Best visibility: October to December (90°N–31°S), but normally at least partially visible from northern latitudes throughout the year

See also the constellations Andromeda, Cassiopeia, Cepheus, Cetus, and Pegasus.

PISCES

The Fishes

Pale with fear, believing her enemy to be close at hand, she held
her child fast in her lap and cried out, "Help me, nymphs!
Come to the rescue of two divinities!" Then, straight away,
she leaped into the Euphrates's waters, and two fishes
swam beneath them. For this service of theirs, the
two fishes—now stars—have their just reward.

Ovid, *Fasti* 2.467–472

MAIN CHARACTERS:
- Aphrodite, Greek goddess of love and desire
- Ares, god of war
- Cupid, the son of Aphrodite by Ares
- Eros, another name for Cupid
- Gaia, the Earth goddess
- Hercules, the greatest of all Greek heroes and famed for his Twelve Labors
- Hydra, a many-headed serpent
- Ladon, the dragon that guarded the apples of the Hesperides

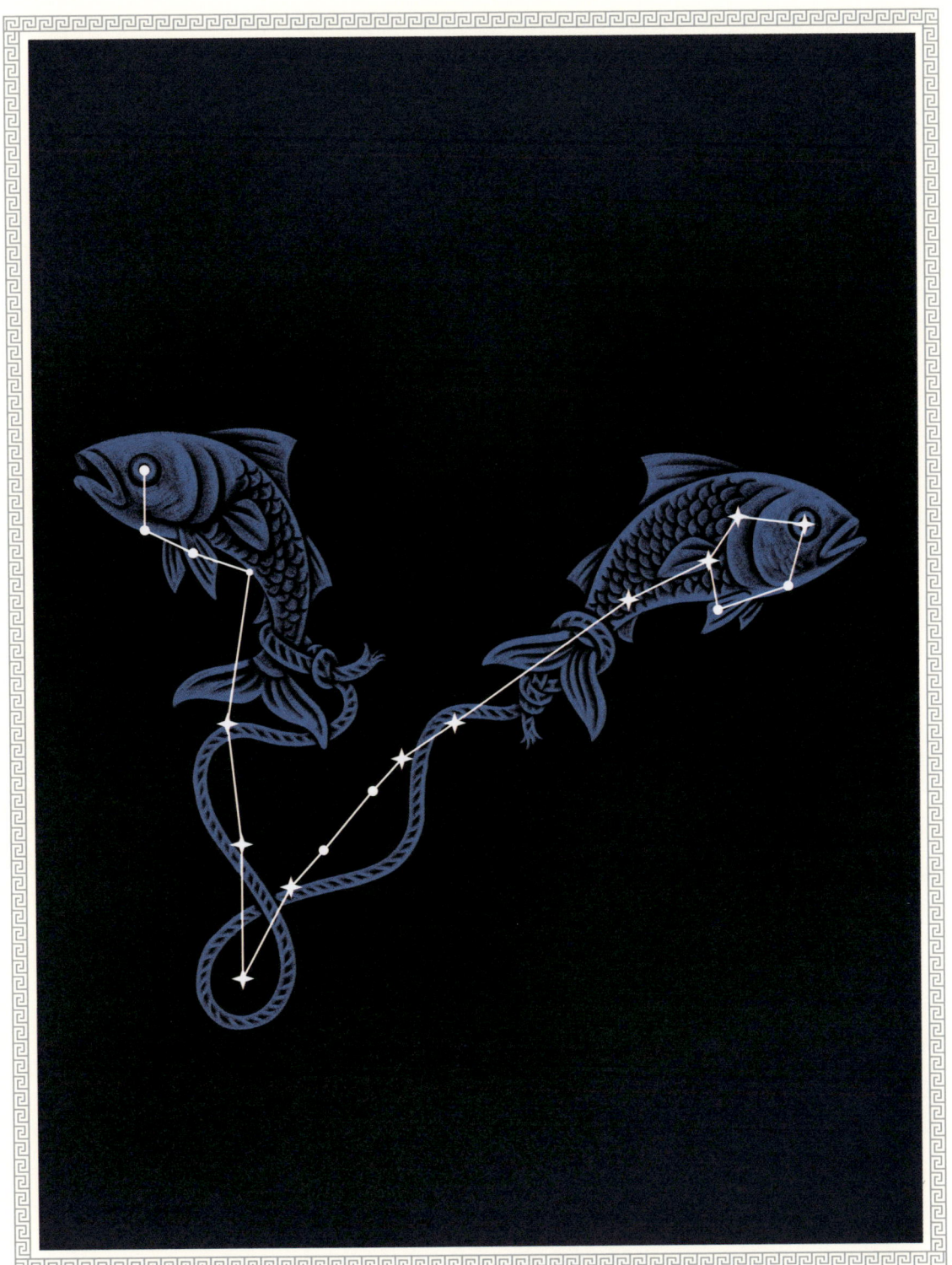

NORTHERN CELESTIAL QUADRANT I
Pisces

CLASSICAL MYTHOLOGY OF THE CONSTELLATIONS

Nemean Lion, a monstrous lion killed by Hercules
Typhon, a hundred-headed monster who waged war on the gods of
   Mount Olympus
Zeus, king of the gods

The two fishes of the constellation Pisces (*the Fishes*) were said to owe their presence in the heavens to Aphrodite, goddess of love and desire. The occasion of their appearance in the night sky was the prideful monster Typhon's attack on Zeus and the other Olympian gods, one by one. Through these assaults, Typhon hoped to become ruler of the universe. Typhon himself was a child of Gaia, goddess of the Earth, and he was fearsome to behold, having one hundred fire-breathing serpent heads. And, as it happened, Typhon was the father of a host of other monsters slain by Hercules and other heroes. Among his children, several of whom became constellations, were the Nemean Lion (Leo), the many-headed Hydra, and the sleepless dragon Ladon (Draco).

When Typhon directed his assault on Aphrodite, she retreated to Syria and, with Typhon in hot pursuit, found herself on the banks of the Euphrates River. She had with her Cupid, or Eros as he was also called, her baby by Ares, the god of war. Aphrodite begged the river's resident nymphs to help her, and this they did. When the desperate goddess leaped into the water to escape, holding Cupid fast, two fishes suddenly appeared, swimming beneath mother and child and carrying them to safety on their backs.

In an alternate version of this tale, Aphrodite and Cupid themselves were transformed, temporarily, into fishes so that they could make their escape. Subsequently, as this story goes, people living in the region ceased thereafter to eat fish, fearing that they might accidentally catch and eat a goddess and her son.

In any event, the fishes were pictured among the stars to commemorate Aphrodite and Cupid's dramatic rescue.

There is, however, yet another version of the Pisces story, and it, too, centers on the goddess Aphrodite. According to this, an egg of unusually large size rolled into the Euphrates River, and some fishes, finding it, pushed it to shore. There it was discovered by a dove, who sat on it until it hatched. What emerged from the egg was no bird but Aphrodite herself. From that time on, as a sign of respect and reverence for those creatures that had saved the goddess, local people ate neither fish nor doves. As for the fishes who rescued her, these were placed in the skies as a constellation: two fishes eternally linked to each other, bound at their tails by a rope.

Best visibility: October to November (83°N–56°S). Pisces is one of the twelve constellations of the Zodiac, with the Sun passing through it from mid-March to mid-April. One of the fishes' heads extends into Northern Quadrant 4.

See also the constellations Draco, Hercules, Hydra, and Leo.

TAURUS

The Bull

> Pasiphaë, in love with the bull, begged Daedalus to create some device to lure that creature; and so he fashioned a hollow cow, just like the cows in the bull's familiar herd. What their union produced was the strangely hybrid creature known as the Minotaur.
>
> ———
>
> Philostratus, *Imagines* 1.16

MAIN CHARACTERS:
Ariadne, a princess of Crete
Daedalus, the legendary craftsman who built the Minotaur's labyrinth
Europa, a princess of Tyre who was kidnapped by Zeus
Hercules, the greatest of all Greek heroes and famed for his Twelve Labors
Minos, king of Crete
Minotaur, half-bull and half-human offspring of Pasiphaë
Pasiphaë, queen of Crete
Poseidon, god of the sea
Theseus, prince of Athens and slayer of the Minotaur
Zeus, king of the Greek gods

In antiquity, there was some confusion about exactly which bull from Greek mythology is represented by the constellation Taurus (*the Bull*), but there were two primary candidates. As it happens, the stories of both had to do with the island of Crete and Minos, that island's king. The most widely accepted notion appears to have been that the constellation's bull is the one that Hercules was sent to capture and bring back alive to Greece as his seventh labor. This splendid bull, the

NORTHERN CELESTIAL QUADRANT I
Taurus

so-called Cretan Bull, had emerged from the sea, a miracle that ensured Minos's claim to the throne. The reign of Crete had been contested, and Minos had prayed to Poseidon, god of the sea, to help him. This Poseidon did by sending the bull. However, Minos was so taken with the animal that he could not bring himself to sacrifice it to Poseidon in thanks. For this failure he would pay a heavy price.

Poseidon then caused Pasiphaë, Minos's wife, to develop an insatiable lust for Minos's bull. In the end, her desire for it was so great that she asked the renowned craftsman Daedalus to come to her assistance, and Daedalus fabricated a life-sized hollow cow that Pasiphaë could climb into. Her coupling with the bull achieved, Pasiphaë fell pregnant and, in the course of time, gave birth to the Minotaur, half-human and half-bull. The name Minotaur means "bull of Minos." Not only was this creature fear-inspiring to look at but it also had a taste for human flesh. Out of shame and horror, Minos asked Daedalus to construct a maze—the famed labyrinth—in which to imprison it.

As for its diet, the Minotaur was periodically fed seven youths and seven maidens sent as tribute to Crete from the subject kingdom of Athens. This continued for a considerable time until the Athenian hero Theseus arrived on the island. With the help of Minos's daughter Ariadne, Theseus slew the Minotaur and emerged safely from the labyrinth. Meanwhile, Poseidon's bull, which was the Minotaur's father, roamed the countryside in Crete until Hercules was sent to capture it and bring it to mainland Greece. This he did, carrying the beast on his shoulders, and eventually set it free at Marathon. There Theseus later killed it as a sacrifice to Athena, patron goddess of Athens, and Zeus placed the bull among the stars along with a host of other animals and monsters who had featured in the Labors of Hercules.

Alternatively, the constellation Taurus reflects the tale of Europa, princess of Tyre in what is now Lebanon. Known for his wandering eye, the god Zeus fell in love with Europa and contrived to find a way to approach her. This he did by transforming himself into a beautiful, tame white bull, and, disguised in this way, appeared before her one day on the seashore. The sight of this bull was amazing to Europa, as was its gentleness. Completely charmed, Europa plaited wreaths of meadow flowers, which she placed upon its head. Then, when the bull bowed down, she dared to climb onto its back. As soon as she did this, Zeus leaped up and plunged into the waves, carrying her all the way to Crete. There, she would become mother to Minos and two other sons by Zeus.

Whether Taurus is the Minotaur's father or the god Zeus in disguise, the constellation depicts only the forequarters (head, front legs, and shoulders) of the bull as it presumably rises from heavenly waves.

Best visibility: December to January (88°N–58°S). Taurus is one of the twelve constellations of the Zodiac, with the Sun passing through it from mid-May to mid-June.

See also the constellation Hercules.

TRIANGULUM

The Triangle

Longing for her vanished daughter, Demeter wandered, searching all the Earth. Now Sicily, all along the slopes of Etna, was covered with streams of volcanic fire, and the land groaned over its whole length; due to a mother's grief over the maiden's loss, the people there, beloved of Zeus, perished for lack of grain.

Diodorus Siculus, *Library of History* 5.5.1

MAIN CHARACTERS:
 Cronus, youngest of the Titan gods and father of Zeus
 Demeter, goddess of grain and the harvest
 Hades, god of the Underworld
 Hermes, the messenger god
 Persephone, daughter of Demeter and wife of Hades
 Poseidon, god of the sea
 Zeus, king of the Greek gods

The small but distinctive constellation Triangulum (*the Triangle*) consists of just three stars, each marking one of the triangle's three points or vertices, as these points are called. This Northern Celestial Hemisphere constellation is to be distinguished from Triangulum Australe (*the Southern Triangle*) which has been known since the Renaissance but was not known in antiquity.

There are a number of myths related to Triangulum, and it can be no coincidence that at least one of these is somehow linked to the Greek letter delta, which is a triangle when written in upper case. The story in question centers on Demeter, goddess of grain and the harvest, and on the island of Sicily, which was a major center of agricultural activity in antiquity. Notably, Demeter's name begins

NORTHERN CELESTIAL QUADRANT I
Triangulum

CLASSICAL MYTHOLOGY OF THE CONSTELLATIONS

with the letter delta, and the shape of Sicily is almost perfectly triangular, or so it was viewed by the Greeks and Romans.

Demeter is best known in mythology as the mother of Persephone, who became the wife of Hades, god of the dead. When Hades was looking for a bride, he knew the task would not be an easy one. No one could possibly want to go live with him in the depths of the Underworld, no matter how grand his palace. For this reason, he decided to take a bride by force. The young woman that he chose was Persephone, daughter of Demeter and Zeus, whom he seized from a meadow where she was picking flowers. By some accounts, that meadow was on the island of Sicily.

A despondent Demeter wandered the Earth searching for Persephone, but to no avail. As a result of her growing desperation and anger, seeds failed to sprout, and crops failed to grow. Crops could thrive only when the goddess of the harvest was content. Sicilian farmers and their livestock, in particular, were hard hit. Eventually, even the gods began to suffer, as people could neither feed themselves nor make offerings to the gods. This prompted Zeus to intervene and approach Hades, who agreed to return Persephone to the world above for a portion of each year. He reasoned that because Persephone had eaten the seeds of a pomegranate in the Underworld, this bound her to that region permanently, at least to some degree. The months when seeds remain dormant in the ground mark the time that Persephone is in the Underworld, but when they sprout and grow in spring and summer, she is with her mother. As a memorial to these events, Demeter pictured Sicily in the heavens as a constellation.

An alternate myth behind the naming and identification of Triangulum has to do with the moment when Zeus ascended to a position of supreme power among the gods. Having defeated his father Cronus and Cronus's siblings, the Titans, Zeus was left king of the gods but saw fit to divide governance of the universe between himself and his siblings Poseidon and Hades. By drawing lots, Zeus was designated lord of the sky, Poseidon lord of the sea, and Hades lord of the Underworld. This threefold division, then, is represented by the triangle's three vertices.

A different tradition explains Triangulum as a reflection of the god Hermes's invention of the Greek alphabet and, in particular, his design of the letter delta, the first letter in some forms of Zeus's name. According to yet another story, Hermes placed Triangulum in the heavens by the head of Aries (*the Ram*) in order to make the ram's head easier to see. Additionally, though not a myth per se, the constellation was believed by some to be a representation of the Nile Delta, the triangular deposition of sediment where the Nile's waters empty into the Mediterranean Sea.

Best visibility: November to December (90°N–52°S)

See also the constellation Aries.

NORT
CELE
QUAD

HERN
STIAL
RANT
II

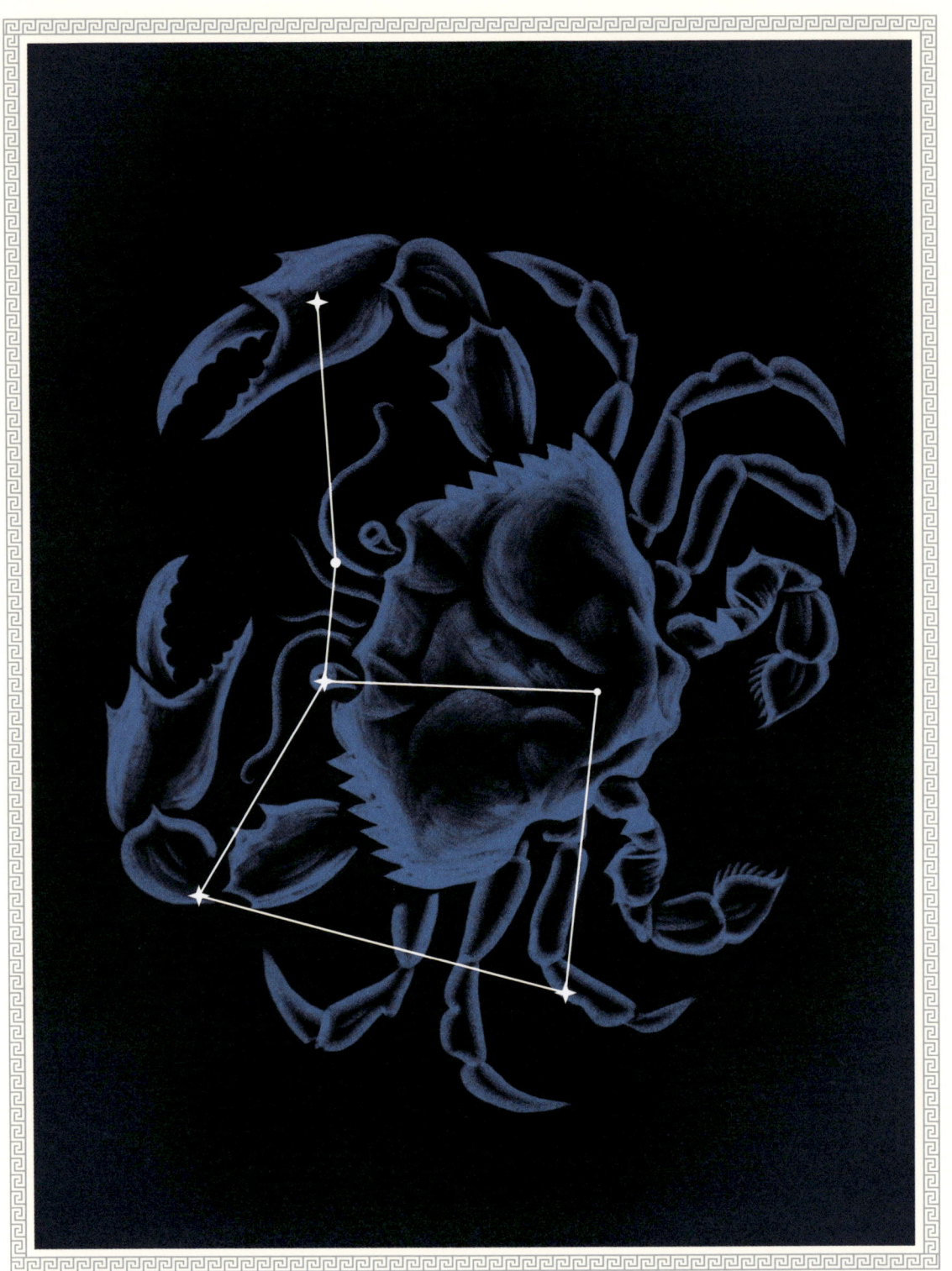

NORTHERN CELESTIAL QUADRANT II

Cancer

NORTHERN CELESTIAL QUADRANT II

CANCER

The Crab

The Crab is said to have been put among the stars by the favor of Hera, because, when Hercules had battled against the Lernaean Hydra, it had snapped at his foot from the swamp.

Hyginus, *Astronomica* 2.23.1

MAIN CHARACTERS:

Cancer, an enormous crab that attacked Hercules
Gemini, the twin brothers of Helen of Troy
Hera, queen of the gods
Hercules, greatest of the Greek heroes and famous for his Twelve Labors
Hydra of Lerna, a many-headed serpent and adversary of Hercules
Leo, the Nemean Lion killed in Hercules's first labor

The constellation Cancer (*the Crab*) is the starry image of a crab that played a relatively minor but interesting role in the hero Hercules's famous labors. One of the twelve near-impossible tasks that the goddess Hera imposed on Hercules was killing the dreaded Hydra of Lerna. The Hydra, a many-headed serpent, inhabited the marshes of Lerna on the eastern coast of the Peloponnese. Hercules battled long and hard with the Hydra, severing each of its heads in turn but discovering, to his horror, that they all quickly grew back, doubling in number. As Hercules struggled with the Hydra, a giant crab emerged from the adjacent water and pinched at Hercules's ankles. In a rage, Hercules stomped the annoying crab to death. The goddess Hera, who regarded the crab as a valiant ally in her ongoing efforts to punish Hercules, rewarded him by placing him among the stars. As a constellation, the crab now resides between the constellations of Leo and Gemini.

Best visibility: February to March (90°N–57°S). Cancer is one of the twelve constellations of the Zodiac, with the Sun passing through it from mid-July to mid-August.

See also the constellations Gemini, Hercules, Hydra, and Leo.

CANIS MINOR

The Lesser Dog

A hill rose from the plains below, and I climbed it so I
could watch this unprecedented race. Now the fox seemed to be
caught, now it escaped unharmed. Skillfully it fled, not straight
but veering suddenly, eluding the jaws of its pursuer. Then it
circled back again to escape its enemy's attack. Yet the hound
still threatens, keeping pace, but with each snap of
his jaws, however close, gets only bites of air.

———

Ovid, *Metamorphoses* 7.779–786

MAIN CHARACTERS:
- Amphitryon, a son of the hero Perseus
- Artemis, goddess of the hunt and protectress of wild animals
- Cephalus, a hunter and friend of Amphitryon
- Creon, a king of Thebes
- Dionysus, the god of wine and the vintage
- Hera, queen of the gods and wife of Zeus
- Icarius, an Athenian farmer who was a devotee of Dionysus
- Laelaps, a hunting dog that always caught its prey
- Maera, Icarius's faithful dog
- Orion, one of classical mythology's greatest hunters
- Teumessian Fox, a fox that terrorized the Teumessians and could not be caught
- Zeus, king of the gods

There was considerable disagreement among ancient authors regarding the identity and mythology of Canis Minor (*the Lesser Dog*), and there was a general confusion, or conflating, of this dog with the dog represented by the constellation Canis Major (*the Greater Dog*).

One animal identified as Canis Minor was no dog at all: the so-called Teumessian Fox. This particular fox had been sent by the goddess Hera (or the god Dionysus) to the town of Teumessus in the territory of the city of Thebes. Its purpose was to terrorize the inhabitants by killing their flocks and even their children, but it is unclear what, exactly, the Teumessians were being punished for.

NORTHERN CELESTIAL QUADRANT II
Canis Minor

CLASSICAL MYTHOLOGY OF THE CONSTELLATIONS

Apart from being a terror, this fox, as the deity that sent it ensured, could never be caught. In the course of time, the hero Amphitryon tried his hand at ridding the region of this fox. The fox's elimination was the condition Creon, king of Thebes, imposed on him: only then would Creon agree to join Amphitryon in avenging the death of his brothers-in-law. At any rate, Amphitryon was well aware that he would need help to catch the fox, so he sought out a young hunter called Cephalus, who had been given Laelaps, a marvelous hunting dog that always caught its prey. When Laelaps was released in pursuit of the fox, the chase went on and on. It would have continued endlessly had Zeus not intervened and stopped the animals cold, turning them to stone. This spectacular chase was then memorialized for all time by the appearance of both the fox and Laelaps among the stars, the fox as Canis Minor and Laelaps as Canis Major.

For some authors, on the other hand, Canis Minor was simply the smaller of Orion's two hunting dogs. For others, Canis Minor was Orion's one and only dog. In any event, Orion was said to have been raised to the stars as a constellation by the goddess Artemis with one, or both, of his hunting dogs after being fatally stung by a scorpion. The scorpion, a creature of monstrous size, had been sent by the gods to kill Orion as punishment for boasting, or threatening, that he could hunt and kill all the animals on the island of Crete. Zeus commemorated the scorpion's role in this event by placing it in the heavens as the constellation Scorpius. Thereupon, the goddess Artemis, who once had been Orion's cherished hunting companion, persuaded Zeus to raise Orion and his dog(s) to the heavens, too, in memory of his heroic feats.

Another dog associated with Canis Minor was Maera, the loyal dog of the devout Athenian farmer Icarius. Maera had led Icarius's daughter Erigone to Icarius's body after he had gone missing, murdered by some neighboring shepherds. Icarius had offered these shepherds wine, a gift from the god Dionysus, but they overindulged in this new drink, becoming stunned and then fearful. They wrongfully suspected Icarius of poisoning them and attacked him with their clubs. Upon finding her father dead, a grief-stricken Erigone hung herself, and the faithful dog Maera, equally despondent, leaped into a well to her death. Zeus was said to have taken pity on this family and placed all three, Maera included, among the stars. Icarius became Boötes, Erigone became Virgo, and Maera became Canis Minor.

Best visibility: February (89°N–77°S)

See also the constellations Boötes, Canis Major, Orion, Scorpius, and Virgo.

NORTHERN CELESTIAL QUADRANT II

GEMINI

The Twins

> Now, in the great hunt for the boar that ravaged Calydonia, the Gemini—not yet stars in heaven—came cantering up, both on snow-white steeds, both prepared to strike, holding aloft spears with brightly flashing points.
>
> Ovid, *Metamorphoses* 8.372–375

MAIN CHARACTERS:

Amycus, barbarous king of the Bebryces
Artemis, protector-goddess of wild animals
Atalanta, a famously swift-footed huntress
Castor, the mortal brother of Pollux and one of Gemini's twins
Helen of Troy, the famously beautiful princess of Sparta abducted by Troy's prince Paris
Hercules, the greatest of all Greek heroes and famous for his Twelve Labors
Iphicles, Hercules's twin brother
Jason, the hero who led the expedition of the ship *Argo* to seek the golden fleece
Leda, a queen of Sparta and mother of Castor and Pollux
Meleager, the hero who killed the Calydonian boar
Pollux, a son of Zeus and twin brother of Castor
Tyndareus, a king of Sparta and father of Castor
Zeus, king of the gods

The constellation Gemini (*the Twins*) memorializes Castor and Pollux, the remarkable twin brothers of Helen of Troy. Like the great hero Hercules and his twin brother Iphicles, Castor and Pollux had different fathers. Pollux was the son of Leda, the queen of Sparta, by Zeus, who seduced her disguised as a swan. This particular transformation on Zeus's part was memorialized by the constellation Cygnus (*the Swan*). Castor, on the other hand, was Leda's son by her mortal husband, King Tyndareus. Consequently, Castor was mortal while Pollux was immortal, though both brothers would achieve immortality at the end of

NORTHERN CELESTIAL QUADRANT II
Gemini

their lives. Notably, their lives were marked not only by heroic exploits but also extraordinary kindness toward humans in peril and need, especially seafarers.

As for their particular skills and talents, Castor was known as an expert tamer of horses and served as one of Hercules's childhood instructors, teaching him both horsemanship and fencing. Pollux, meanwhile, was especially good at boxing, a skill that he used to good effect in one of their first adventures, the expedition of the ship *Argo* to the distant shores of the Black Sea. On the *Argo*, the twins joined Jason and some of the best and bravest men of Greece in their perilous quest for the golden fleece of a very special ram (Aries). The *Argo*'s crew were known as the Argonauts, a name that means "sailors on the *Argo*." Among the many adversaries that the Argonauts faced was the barbarous king of a tribe called the Bebryces, who were said to have lived in what is now northwest Turkey. This king, Amycus, challenged all newcomers to his territory to a boxing match instead of offering them hospitality. In the Greek world, hospitality toward strangers was the norm and, indeed, a hallmark of Greek culture, so the challenge was an unexpected and alarming affront. Until the arrival of the Argonauts, Amycus had been undefeated, but with Pollux he finally met his match. The latter killed him with a single blow.

The brothers also joined in the famous hunt for the enormous boar that had been sent to ravage the countryside of Calydon in western Greece, where it destroyed crops and killed humans together with their flocks. Artemis, goddess of the wild and protectress of wild animals, had sent the boar as punishment of Calydon's king. A fatal error, that king had forgotten to include Artemis in his annual harvest sacrifices to the gods. The boar was ultimately wounded by swift-footed Atalanta, an expert huntress, and slain by the hero Meleager.

On another occasion, the twins rescued their sister Helen from the Athenian prince Theseus, who had attempted to kidnap her and make her his bride. Strangely, it was Castor and Pollux's own ill-conceived plan to find wives that led to the end of their earthly lives. The women that they fixated on were their cousins, who happened already to have been promised in marriage to others. Ignoring this fact, the twins carried their cousins off to Sparta, doing this on the very day of the women's wedding. Inevitably, hostilities between the twins and their brides' intended grooms ensued. Castor, the mortal brother, was killed in this conflict. A heartbroken Pollux then begged his father, Zeus, to find a means by which he could share his immortality with his brother. Both were subsequently raised to the stars as part of the constellation Gemini. Castor and Pollux are the names of the two brightest stars in this constellation, and although Pollux is brighter, it is Castor that can be seen with the naked eye.

As a further honor to the twins, Zeus made them guardians of seafarers endangered by storms. To those sailors, they appeared as the spectacular atmospheric phenomenon now known as St. Elmo's fire.

Best visibility: January to February (90°N–55°S). Gemini is one of the twelve constellations of the Zodiac, with the Sun passing through it from mid-June to mid-July.

See also the constellations Argo Navis, Aries, Cygnus, and Hercules.

LEO

The Lion

So huge a monster, I am sure, you could not find in all of Argos,
no matter how hard you tried, for there are no other lions
of such size, only bears and boars and wolves.

Theocritus, *Idylls* 25.184–186

MAIN CHARACTERS:

- Cerberus, triple-headed watchdog of the Underworld
- Echidna, the Nemean Lion's half-serpent and half-human mother
- Eurystheus, the evil and cowardly king of Tiryns and Mycenae
- Hera, queen of the gods and wife of Zeus
- Hercules, greatest hero of the Greeks who was tasked with slaying the Nemean Lion
- Hydra of Lerna, a monstrous serpent with many heads
- Nemean Lion, a lion with impenetrable skin who became the constellation Leo
- Typhon, the Nemean Lion's monstrous father with one hundred flaming snake heads
- Zeus, king of the Greek gods and Hercules's father

Leo (*the Lion*) is a constellation commemorating an especially large, fierce lion that terrorized those dwelling in the lands surrounding the sanctuary of the god Zeus at Nemea in the northeastern Peloponnese. The Nemean Lion, as it was called, was one of the monstrous children of Typhon and Echidna, both of them ancient, dread-inspiring creatures born from the Earth goddess Gaia. Typhon,

NORTHERN CELESTIAL QUADRANT II
Leo

the lion's father, had one hundred flaming snake heads that spoke with human and animal voices. Echidna, Typhon's mate, was half-snake and half-woman. The many-headed Hydra and the hellhound Cerberus were among the lion's gruesome siblings. Others were Ladon, the dragon (of the constellation Draco) that guarded the Garden of the Hesperides, and the Chimera, a monster that was part lion, part goat, and part dragon.

Given the lion's massive size and ferocious disposition, the lion was an ideal means by which the goddess Hera could avenge herself on Hercules. As was often the case in classical mythology, a human punished by the gods was not necessarily guilty of any deliberate offense. In this instance, Hercules happened to be the illegitimate child of Hera's husband, Zeus, and she was fed up with Zeus's infidelity. As a result, she became determined to ensure that Hercules would have a life of hardship. It was she who set into motion the sequence of events leading directly to Hercules's famous Twelve Labors.

The first labor that Hercules was compelled to complete was to kill the Nemean Lion and present it to Eurystheus, king of the powerful kingdom of Tiryns. When Hercules arrived at Nemea and caught sight of his prey, he took aim with his bow and arrow. His aim was true, but the arrow bounced harmlessly from the lion's back. The lion was said to have skin that could not be pierced by any weapon, and Hercules quickly discovered this to be true. Another plan was clearly called for. Hercules observed that, after hunting, the lion retreated daily to its lair, a two-mouthed cave. Waiting till the lion was gone, he sealed one of the cave's openings with a boulder and lay in wait for the beast's return. His new plan was to kill the lion relying on nothing but his strength and skill at wrestling. Cornering the lion, Hercules made his move. He swiftly grabbed its neck and squeezed tight, bringing the beast to its knees. Hercules then made his way back to Tiryns with his lifeless prey slung over his shoulders. After reaching his destination and displaying the lion to King Eurystheus, Hercules stripped off its pelt and wore it over his shoulders like a cloak. From that day forward, Hercules wore this special garment, a hard-won prize and a fitting symbol of his superhuman status. Meanwhile, as means of honoring its loyal service to her, the goddess Hera placed the Nemean Lion's effigy among the stars, depicting it crouching slightly, ready to pounce.

Best visibility: March to April (82°N–57°S). Leo is one of the twelve constellations of the Zodiac, with the Sun passing through it from mid-August to mid-September.

See also the constellations Draco, Hercules, and Hydra.

NORTHERN CELESTIAL QUADRANT II

URSA MAJOR

The Greater Bear

> The goddess grabbed her rival, Callisto, by the hair and dragged her to the ground. Pleading, Callisto raised her arms begging for mercy. But even as she begged, black hair spread over her white limbs, her hands elongated into feet tipped with curving claws, and snarling jaws deformed the mouth that Zeus had kissed.
>
> ———
>
> Ovid, *Metamorphoses* 2.476–481

MAIN CHARACTERS:
- Arcas, the son of Callisto by Zeus
- Artemis, goddess of the wild and protectress of wild animals
- Callisto, a devotee of Artemis who became the mother of Arcas
- Hera, queen of the gods and jealous wife of Zeus
- Lycaon, a king of Arcadia and father of Callisto
- Zeus, king of the gods

The bear who is most often linked to the constellation Ursa Major (*the Greater Bear*) did not begin her life as an animal. Instead, she had been a lovely maiden and the daughter of Lycaon, a king of Arcadia in the mountains of the Peloponnese. Callisto, whose name means "the fairest," was as chaste and pure as she was beautiful, and she was a devoted follower of the virgin goddess Artemis. On one fateful day, the god Zeus caught sight of Callisto and was determined to make her his own. Knowing that it would be difficult to win the heart of someone so pure, he did an unspeakable thing: he changed his appearance to resemble Artemis. Sure enough, Callisto did not fear him as he approached in this goddess's form, but she was soon terrified and resisted strenuously when he did reveal himself. Being a mere mortal, however, she could not hope to escape the powerful god's shameful and despicable assault.

In the course of time, Callisto's belly began to grow. For a while, she was able to hide this from her beloved Artemis until the goddess and the other maidens in her company found a secluded pool in which to bathe. Uncharacteristically, Callisto refused to join the others, and her secret was soon revealed. The angry goddess then sent her away, far from her home and companions. Worse still, once her child

CLASSICAL MYTHOLOGY OF THE CONSTELLATIONS

by Zeus was born, the goddess Hera took her vengeance on the poor girl, too. Enraged that her husband, Zeus, had strayed—a thing that he too often did—Hera changed Callisto into a bear. It was as a bear that Callisto now roamed Arcadia, and as a bear she eventually came upon her son, Arcas, fully grown and out in the wilderness to hunt. Arcas raised his spear and would have killed her if Zeus had not intervened, raising them to the heavens as constellations, both appearing there as bears. Arcas now strides through the night sky alongside Callisto as the Lesser Bear, Ursa Minor.

A more obscure myth offers an alternate explanation for the heavenly bears. This story, too, features the god Zeus but centers on his childhood. As a baby, Zeus had been raised by nymphs who lived in a cave on Mount Ida on the island of Crete. Zeus had been sent there by his mother, the Titan goddess Rhea, who feared for his life—for good reason, as her husband, Cronus, had just eaten her other five newborn children. As a reward for keeping him safe, Zeus placed his surrogate mothers in the sky as the constellations Ursa Major and Ursa Minor.

The constellation Ursa Major extends slightly into Northern Quadrant 3, this being where the bear's tail is located, and famously contains the seven-star asterism (a distinctive but informally recognized pattern of stars) known in North America as the Big Dipper, one of the most easily recognizable patterns in the sky.

Best visibility: February to May (90°N–16°S)

See also the constellation Ursa Minor.

NORTHERN CELESTIAL QUADRANT II
Ursa Major

NORT
CELE
QUAD

HERN STIAL RANT III

NORTHERN CELESTIAL QUADRANT III
Boötes

NORTHERN CELESTIAL QUADRANT III

BOÖTES

The Herdsman

Having received the god's precious gift, Icarius straight away loaded his ox-drawn wagon with skins filled with wine. For this he is known as Boötes. But as he made his way through the Attic countryside and showed it to the shepherds, then some of them, filled with greed for this drink newly introduced to them, fell into a stupor, their limbs sprawled everywhere, as if half dead.

Hyginus, *Astronomica* 2.4.3

MAIN CHARACTERS:
- Arcas, the son of Callisto by Zeus
- Callisto, a young woman loved by Zeus and transformed into a bear
- Demeter, goddess of grain and the harvest
- Dionysus, the vastly popular god of vegetation and wine
- Erigone, the devoted daughter of the noble Icarius
- Hera, queen of the gods and jealous wife of Zeus
- Iasion, the father of twin sons with Demeter
- Icarius, a devout and generous Athenian farmer
- Maera, Icarius's devoted dog
- Philomelus, a son of Demeter and inventor of the oxcart
- Plutus, a god of Earth's buried riches
- Zeus, king of the Greek gods and father of Arcas

The name Boötes, which is typically translated as "the Herdsman," literally means "ox herder" or "ox driver" in ancient Greek. There are a number of myths associated with the large constellation Boötes, some more obscure than others and some not involving cattle at all.

Among those figures identified as Boötes is an Athenian man by the name of Icarius. He and his daughter Erigone were notable for welcoming the god Dionysus when others, fearful of this foreign god, were less than eager to do so. Dionysus was a god of vegetation and of wine. He was also a democratic god in whose eyes everyone was equal: young and old, male and female, enslaved and free, even animal and human. To some, the appearance of this gender-fluid god was

CLASSICAL MYTHOLOGY OF THE CONSTELLATIONS

troubling, as was his ability to relieve those who believed in him from their worries and the constraints of their daily lives. As a reward for his devotion, Dionysus taught Icarius the valuable art of winemaking, and the grateful, generous man then loaded his oxcart with wineskins to share this gift and his new knowledge with his neighbors.

Icarius first came upon a group of shepherds, who drank heartily of this new beverage, quickly becoming drunk and then violently ill—so ill that they believed Icarius had poisoned them. In their stupefied, addled state, they attacked poor Icarius, bludgeoned him to death, then ran away. As her father was gone longer than expected, a worried Erigone set out to look for him and, at long last, was led to his body by Maera, Icarius's faithful dog, who had accompanied him on his fateful errand. The maiden Erigone was so despondent over her father's death that she hung herself from a tree, and the heartbroken dog Maera leaped into a well to her death. All three were depicted in the heavens by Zeus, who acted out of pity for their terrible fate. Icarius was shown as the constellation Boötes, and the maiden Erigone was pictured in the constellation Virgo. Faithful Maera's image was claimed by some to be reflected in the Dog Star, Canicula ("little dog," also called Sirius), the brightest star visible from Earth in the night sky and of enormous cultural significance in antiquity. Others claimed that Maera became one of the two dog constellations Canis Major and Minor (*the Greater Dog* and *the Lesser Dog*).

By some accounts, Boötes is instead a son of Demeter, goddess of grain and the harvest. Though a goddess, Demeter fell in love with a mortal man called Iasion and bore him twin sons. One was Plutus, who became a deity of Earth's stored wealth, and the other was Philomelus, who became a humble farmer. Poor and unassuming though he was, he brought his mother great joy by inventing the ox-driven plow, a vital instrument for all later farmers. A proud Demeter honored him by representing him at the plow in the heavens.

Another figure sometimes identified with Boötes is Arcas, the son of Callisto. Those who believed that the constellation was Arcas also called it by a different name, Arctophylax, which means "bear's guardian." The bear in question is Callisto, a young woman who had been raped by the god Zeus and impregnated by him. Even though Callisto had not lain with Zeus willingly, a jealous Hera, fed up with Zeus's many infidelities, transformed her into a bear as punishment once her child had been born. When grown, that child, Arcas, would come very close to killing his own mother, having no idea who the bear appearing suddenly before him in the forest actually was. Tragedy was prevented by Zeus, who placed mother and son among the stars, Callisto as Ursa Major (*the Greater Bear*) and Arcas as Arctophylax. This particular version of Callisto's tale runs contrary to that in which mother and son both become starry bears, and Arcas is not Arctophylax but rather Ursa Minor (*the Lesser Bear*).

Because identifications of this constellation varied, depictions of it are varied as well. Generally, Boötes is shown holding up a herdsman's staff in his right hand. In his left, he is sometimes shown holding a small sickle and sometimes holding the leashes of two hunting dogs. These dogs are associated with the postclassical constellation Canes Venatici (*the Hunting Dogs*).

Best visibility: May to June (90°N–35°S)

See also the constellations Canes Venatici, Canis Major, Canis Minor, Ursa Major, Ursa Minor, and Virgo.

CORONA BOREALIS

The Northern Crown

Theseus sailed away—pitiless!—abandoning the princess Ariadne deep in slumber on the shore, scattering his worthless promises to the winds. For him she had betrayed her country, risking everything. But Dionysus caught sight of deserted Ariadne sleeping and was overcome with love and wonder.

Nonnus, *Dionysiaca* 47.269–273

MAIN CHARACTERS:
- Aphrodite, Greek goddess of love and beauty
- Ariadne, a princess of Crete who wed Dionysus
- Dionysus, god of wine, vegetation, and the theater
- Minos, a legendary king of Crete
- Minotaur, a hybrid human-bull monster
- Pasiphaë, queen of Crete and mother of the Minotaur
- Poseidon, god of the sea
- Theseus, a prince of Athens and slayer of the Minotaur

The constellation Corona Borealis, literally "northern crown," is the crown-shaped constellation in the skies of the Northern Celestial Hemisphere. In the original Greek text of Ptolemy's *Almagest*, its name is simply Stephanos, which translates in English to "wreath" or "crown." In antiquity, wreaths *were* crowns, and they were made of olive branches, laurel, and other plants, sometimes re-created in gold. The adjective *borealis* ("northern") was a later addition, distinguishing this crown from that in the Southern Celestial Hemisphere (Corona Australis).

Corona Borealis was generally believed to represent the crown of the princess Ariadne. She was the daughter of Minos, king of Crete, and of Pasiphaë, Crete's queen. Pasiphaë is best known for the unnatural passion she developed for a bull that had been sent to Crete by the god Poseidon. The bull's presence, together with the harm and heartbreak it caused, was intended as punishment for Minos because he had not appropriately thanked Poseidon for supporting his bid for kingship. Impregnated by the bull, Pasiphaë became mother to the dreaded Minotaur, part human and part bull, who fed on human flesh.

Out of shame mixed with horror, King Minos kept his monstrous stepson, the Minotaur, imprisoned in a labyrinth. As for the creature's food, this was supplied by the Athenians, who had been defeated in battle by Minos and were thereafter bound to send seven youths and seven maidens to Crete every nine years as tribute. When Theseus, the prince of Athens, reached maturity, he volunteered to go along with the thirteen others who had been chosen by lot. His intention was to free Athens once and for all from Minos's barbarous oppression.

When the Athenians arrived on the island of Crete, it was love at first sight for Ariadne. She now had a terrible decision to make: should she remain loyal to her family and country, or should she help Theseus and the Athenians? Her heart won out. Killing the Minotaur was within Theseus's grasp, but finding his way out of the labyrinth would be near impossible. It was Ariadne who came to the rescue, giving Theseus a ball of yarn to unroll while making his way into the heart of the labyrinth and to follow on his way out. In gratitude, Theseus agreed to take her back with him to Athens as his bride. Having betrayed her family, she could not possibly stay. The Minotaur was duly slain, and Theseus, with Ariadne, made his escape.

What Theseus did next was unforgivable. On the return voyage to Athens, Theseus instructed his crew to moor at the island of Naxos for the night, or so Ariadne was led to believe. However, while she slept on the island, Theseus and his crew sailed away, abandoning her all alone. It was her good fortune that the god Dionysus happened to catch sight of her while she slept and carried her away to be his bride. Corona Borealis represents the crown of gold and precious gems from India given to Ariadne by the goddess Aphrodite as a wedding gift. Upon her

NORTHERN CELESTIAL QUADRANT III
Corona Borealis

death, the gods placed this crown in the heavens as a tribute to her. The Minotaur, meanwhile, makes his appearance in the constellation Taurus (*the Bull*).

Best visibility: June (90°N–50°S)

See also the constellations Corona Australis and Taurus.

DRACO

The Dragon

Now, slain by Hercules' arrows, the dragon of the Hesperides lay
prostrate beside the apple tree. The tip of his tail still twitched,
but from the head down, all along his dark spine, he showed no
other sign of life. In his pooling blood, poisoned by arrows that
had been dipped in the Hydra's lethal gall, flies perished
as they swarmed the festering wounds.

Apollonius of Rhodes, *Argonautica* 4.1400–1405

MAIN CHARACTERS:
Atlas, the Titan god who held up the heavens on his shoulders
Cerberus, the triple-headed watchdog of the Underworld
Echidna, monstrous mother of Ladon
Gaia, Greek goddess of the Earth
Hera, queen of the Greek gods and wife of Zeus
Hercules, greatest of the Greek heroes
Hesperides, nymphs of the Far West
Ladon, the dragon that guarded the Garden of the Hesperides
Nemean Lion, a monstrous lion killed by Hercules
Typhon, the monstrous father of Ladon
Zeus, king of the Greek gods

Ladon, the dragon tasked with guarding the Garden of the Hesperides, is the namesake of the constellation Draco, "the dragon" in Greek. The garden that he watched over was a grove of apple trees given by the Earth goddess Gaia to Hera, queen of the Greek gods, on the day of her wedding to Zeus. The trees were tended by the nymphs known as the Hesperides, who lived in the Far West

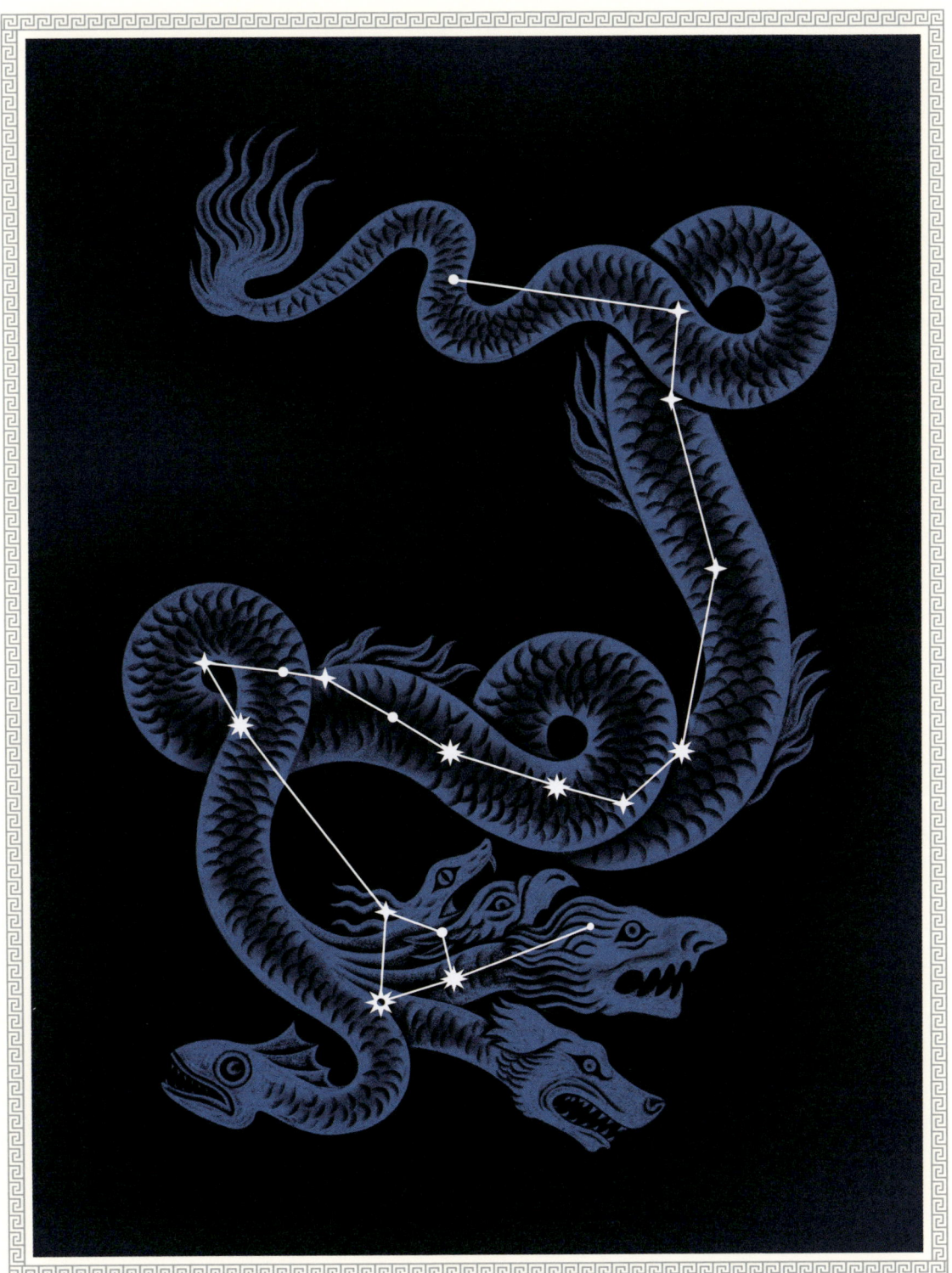

NORTHERN CELESTIAL QUADRANT III

Draco

(of the world known to the Greeks) in what some authors identified as the Atlas Mountains region of modern Morocco. Notably, the apples that these precious trees bore were golden, and they were also a source of eternal life.

No ordinary dragon, Ladon came from an infamous family of monsters. His parents were ancient hybrid creatures, Typhon and Echidna, both of whom were part serpent, and his siblings were the Nemean Lion (Leo), the triple-headed dog Cerberus, the many-headed Hydra, and the snake-lion-goat monstrosity called Chimera. Like his father, Typhon, the dragon Ladon was said to have as many as one hundred heads and to be able to speak in the voice of any creature. Having many heads ensured that some of his eyes were always open and vigilant, and for this reason he was well suited as guardian of the apple grove.

Like the Nemean Lion, the dreaded Hydra, and Cerberus, Ladon encountered Hercules in the course of his famous Twelve Labors. Hercules had been sent to slay the lion and Hydra, but Cerberus was captured alive. In the case of Ladon, the hero needed to evade him, escaping his ever-watchful gaze. His task on this occasion—completion of his eleventh labor—was to pick and take away some of the Hesperides' apples. According to one version of the story, Hercules accomplished the task by killing Ladon with one of his poisoned arrows and then gathering some of the unguarded apples. In an alternate version, rather than attempting to kill Ladon, Hercules tried a different tactic. He approached the Titan god Atlas, who lived nearby, and asked him to collect the apples. As Ladon's familiar neighbor, Atlas could achieve this without great difficulty. So, Hercules offered to take the burden of the heavens, from Atlas's shoulders temporarily, and Atlas went off to retrieve the apples. Whether outwitted by Hercules or slain by him, the goddess Hera still rewarded the dragon for his loyal service and raised him to the heavens, where he joined the Nemean Lion and Hydra's starry forms as the large constellation Draco. Above his head—unnervingly!—treads Hercules himself.

Located largely in Northern Quadrant 3, Draco's neck loops up into Northern Quadrant 4 and his tail extends into Northern Quadrant 2.

Best visibility: March to September (90°N–4°S)

See also the constellations Hercules, Hydra, and Leo.

NORTHERN CELESTIAL QUADRANT III

HERCULES

Hercules

Every sort of terrifying thing the hostile Earth produced, or sea or air brought forth—however frightening, monstrous, poisonous, dreadful, savage—has been broken and tamed by him. He always prevails, growing stronger with every hardship he faces . . . . Now, throughout the whole world, he is spoken of as a god.

Seneca, *Hercules Furens* 30–41

MAIN CHARACTERS:
Alcmene, princess of Mycenae and Tiryns as well as mother of Hercules
Amphitryon, husband of Alcmene and stepfather of Hercules
Apollo, god of prophecy
Artemis, patron goddess of wild animals
Augeas, king of Elis and owner of an enormous herd of cattle
Castor, one of the Gemini and Hercules's instructor
Cerberus, triple-headed watchdog of the Underworld
Ceryneian Hind, a sacred, golden-antlered deer
Chiron, a gentle centaur and Hercules's teacher
Chrysaor, brother of Pegasus and a monstrous child of Medusa
Cretan Bull, sire of the dreaded Minotaur
Deianeira, Hercules's last wife
Erymanthian Boar, a wild boar that roamed Mount Erymanthus
Eurystheus, the evil and cowardly king of Tiryns and Mycenae
Gemini, twin brothers of Helen of Troy
Geryon, the triple-bodied king of Erythrea
Hades, king of the Underworld
Hera, queen of the gods and wife of Zeus
Hercules, a son of Zeus and the greatest of all Greek heroes
Hesperides, nymphs who tended Hera's sacred apple trees
Hesperus, a deity personifying the Evening Star and father (or grandfather) of the Hesperides
Hippolyta, queen of the Amazons
Hydra of Lerna, a many-headed serpent

CLASSICAL MYTHOLOGY OF THE CONSTELLATIONS

Hylas, Hercules's young companion who was kidnapped by water nymphs
Iolaus, son of Hercules's twin brother, Iphicles
Iphicles, Hercules's twin brother
Ladon, the dragon that guarded the apples of the Hesperides
Mares of Diomedes, flesh-eating horses
Medusa, a snake-haired Gorgon
Megara, first wife of Hercules
Nemean Lion, a lion with impenetrable skin
Nessus, the centaur who would cause Hercules's death
Pegasus, the famous winged horse
Persephone, queen of the Underworld
Philoctetes, the compassionate man who helped Hercules end his life
Stymphalian Birds, birds that shot their feathers like arrows
Zeus, king of the Greek gods and Hercules's father

It is only fitting that Hercules, the best-known hero of Greek mythology, found a place among the stars. Although he was universally loved and revered in antiquity, being regarded as a savior of humankind as well as a paragon of bravery, his was a life of hardship and tragedy from the moment of his birth. The source of his suffering was the goddess Hera, queen of the gods. She had hoped that the trials to which she subjected him would destroy him; instead, his superhuman endurance raised him to the level of the gods. Ironically, his name in Greek means "Hera's glory," and what follows is the great hero's story.

Hercules's mother was Alcmene, daughter of the king of Mycenae and Tiryns, and she was married to her cousin Amphitryon. Both Alcmene and Amphitryon, incidentally, were grandchildren of the hero Perseus, famed for slaying the Gorgon Medusa. Although Amphitryon was Alcmene's husband, he was not Hercules's biological father. His father, instead, was none other than Zeus, king of the gods. One night, Zeus visited Alcmene in disguise, posing as her husband, so she was surprised and confused when Amphitryon, who had been away at war, returned home the next day expecting an enthusiastic greeting. As it happened, both Zeus and Amphitryon had slept with Alcmene, and she became pregnant with twins. One was a child of a god and the other of a mortal.

As Alcmene's due date drew near, Zeus announced to the assembled gods that a child descended from him would be born on that very day and that this child would rule over Greece's greatest kingdoms. Incensed at this boast and at her husband's infidelity, the goddess Hera planned her revenge. She sent the goddess of childbirth to delay Alcmene's labor and to hasten along the birth of another child instead. That child was Eurystheus, who would become king of Mycenae and of Tiryns, the two most powerful kingdoms in southern Greece. Still, Hera remained angry, and Hercules's delayed birth was only the first stage of her revenge.

Soon afterward the goddess sent a serpent to kill Hercules and his twin brother, Iphicles, while they were asleep in their crib. Iphicles was startled and afraid when the serpent appeared, but Hercules, demonstrating superhuman strength and bravery even as an infant, leaped from the crib and strangled the serpent. Until that time, Alcmene did not know which child had been fathered by Zeus, but now it was clear to everyone. By some accounts, Zeus, wishing to punish Hera, secretly placed Hercules in her arms to nurse while she slept. When she suddenly awoke, the milk flowing from her breast became the Milky Way.

As he matured from child to man, Hercules learned arts and skills typical of Greek education. He learned archery, wrestling, and the use of a sword. His stepfather, Amphitryon, taught him how to drive a chariot; Linus, the brother of the famed musician Orpheus, taught him how to play the lyre (*lyra* in Latin); and Castor, one of the heavenly twins called the Gemini, taught him horsemanship and fencing. Even Chiron, the gentle and highly cultured centaur, was counted in the number of his teachers. All these skills would serve Hercules well in the many harrowing adventures to come. Among the first were a series of wars, and for his efforts in those conflicts he was awarded Megara, daughter of the king of Thebes, as his bride. With Megara, he would become a father, and the family would make their home in Thebes.

As fate would have it, Hercules's family was destined for a tragic end. Again, the cause of misery was Hera. In a sudden, divinely inspired fit of madness, Hercules killed his wife and children, in his crazed state mistaking them for enemies or intruders. When he came to his senses, a devastated Hercules made his way to the god Apollo's oracle at Delphi to inquire how he could atone for his unspeakable—if wholly accidental—crime. At Delphi, Hercules was told that he needed to go to the kingdom of Tiryns and place himself in the service of Eurystheus, the very man whom Hera had made that realm's king by causing him to be born before Hercules. Eurystheus, a cruel and cowardly man, set Hercules a series of tasks, one after another, each time fully expecting to learn that he had not survived. These tasks, twelve in all, came to be known as the Labors of Hercules, and his successful completion of them ensured his enduring fame.

The first of these labors was killing a particular lion living in the wilds of Nemea in the southeastern Peloponnese, where it roamed ravaging the surrounding villagers and their flocks. This lion was a monster born of a monstrous mother,

NORTHERN CELESTIAL QUADRANT III
Hercules

half-woman and half-snake, and its skin could not be penetrated. For all intents and purposes, the lion seemed impossible to kill. Hercules came armed with a bow and arrow, as any hunter would, but his perfectly aimed arrows bounced right off the lion's back. Realizing that a wholly different strategy was required, Hercules hatched a clever plan. He waited inside the lion's lair until it returned at night and sealed its entrance, so the lion could not escape. Using nothing but his strength and skill at wrestling, Hercules choked the life out of the lion and then carried its lifeless form to Tiryns, where he intended to present it to King Eurystheus. But the king, having caught sight of Hercules returning with the beast slung over his back, became so afraid that he ran for cover. Hidden in an enormous storage vessel, Eurystheus made it known that he did not want the lion and that Hercules should not, in the future, enter the palace but rather display proof of his labors' completion outside the city gates. Then Hercules skinned the lion and from that day forward draped the lion's pelt over his head and shoulders. The lion, like Hercules, would become a constellation, Leo, which is Latin for "lion."

For his second labor, Hercules was faced with slaying the lion's gruesome sibling, the Hydra, an enormous many-headed serpent. The Hydra lived on swampy ground outside the region of Lerna, and it was there that Hercules now traveled. Although he was always armed with a club, Hercules's weapon of choice for this task was a sword, his plan being to lop off the Hydra's heads. What he did not know until he tried was that each severed neck sprouted two fresh ones. This was bad enough, but as Hercules hacked at the Hydra's heads, a giant crab emerged from the swamp and pinched his feet. The crab was quickly eliminated, but the doubling heads called for a new strategy. Taking a burning torch from his nephew Iolaus, who had accompanied him on this adventure, Hercules reverted to severing necks, now cauterizing the stumps with his torch so that no new heads would grow. One of the heads proved more problematic still, as it was immortal. This head withstood bludgeoning by Hercules's club and, in the end, the hero buried it beneath a massive boulder. As for the Hydra's body, Hercules cut it open and dipped his arrows in the monster's poisonous gall, blood, and venom. Eventually, the Hydra and the crab (Cancer), too, would become constellations, as would Hercules's arrows (Sagitta), all raised to the heavens by the gods.

Next Hercules was tasked with capturing a very special deer alive. This was the so-called Ceryneian Hind, which was notable for having golden antlers and was sacred to Artemis, patron goddess of wild animals. Weapons clearly were not called for, and for a full year Hercules patiently tracked it. At last, he found the right moment and captured the deer with a net as it slept. He then set off for Tiryns, carrying the unharmed animal over his shoulders. On his way, Hercules came upon Artemis, who was angered over the capture of her sacred animal. But the goddess let him carry on unpunished, considering that it was not Hercules but Eurystheus who was to blame.

Hercules's fourth labor consisted of capturing another live animal, the Erymanthian Boar. This huge, shaggy boar roamed and ravaged the lands around Mount Erymanthus in western Greece. On and around Erymanthus, Hercules tracked it. At last, when winter came, the boar became stuck in the deep snow covering Erymanthus's slopes. It was then that Hercules grabbed it and carried it back to Tiryns as proof of the task's completion. Before capturing the boar, however, Hercules had a series of strange adventures. In the course of his wanderings, he was offered hospitality by a kindly centaur, Pholus, who then was fatally wounded while handling one of Hercules's poisoned arrows. A similar fate befell the centaur Chiron, Hercules's childhood teacher. Both centaurs have been credited with becoming the constellation Centaurus. Hercules also came upon the hero Jason, who had set sail on the ship *Argo* from Greece to the distant shores of the Black Sea. Jason was accompanied by all the bravest men in Greece, their mission being to retrieve the golden fleece of a very special ram (Aries). Hercules wanted to be in their number and, for a time, joined that expedition. He left Jason and the others to search for a young friend, named Hylas, who had become lost and later kidnapped by some water nymphs. In the course of time, the good ship *Argo*, too, was memorialized by the gods as a constellation.

Cleaning out the massive stables of Augeas was Hercules's fifth labor and a difficult, demeaning task. Augeas was the wealthy king of Elis, a town near the famed sanctuary of Zeus at Olympia. The source of Augeas's wealth was an enormous herd of cattle, and their dung had been deposited in the stables so thickly that it defied belief . . . and any normal human capacity to remove it. A man of brains and brawn, Hercules began to dig along the banks of the rivers Alpheus and Peneus to divert their water through Augeas's stables. The rivers' strong currents swept the stables clean.

Hercules's sixth task entailed driving a particularly frightening flock of birds away from Lake Stymphalus in the central Peloponnese. These birds, which were as large as cranes, could shoot their feathers like arrows and had taken refuge at the lake in very large numbers. To rouse them from their roosts, Hercules shook a rattle, and at its sound all the birds took flight. Relying on his skill in archery, Hercules shot some of the birds. The rest were frightened off.

The island of Crete was the site of Hercules's seventh labor, and Minos was that island's legendary king. What Hercules was told to do was capture the infamous bull that had fathered a fearsome, man-eating monster. That monster

NORTHERN CELESTIAL QUADRANT III

was the Minotaur, part human and part bull. Its mother—a source of mortification for Minos—was his own wife and queen. With Minos's help, Hercules rounded up the bull and took it to mainland Greece, where he set it free to roam after displaying it to Eurystheus. The Cretan Bull (Taurus) would become a constellation, too.

For Hercules's eighth task, Eurystheus sent him to Thrace in northern Greece to fetch the terrifying Mares of Diomedes. These mares fed on human flesh and belonged to the savage king of a people called the Bistones. On his way to Thrace, Hercules stopped to help Admetus, the king of Pherae, whose wife Alcestis had died a premature death. Hercules wrestled with Hades, god of death, and returned Alcestis to the living. As for the mares, he rounded them up and fed them the flesh of Diomedes, their own master, which in turn served to make them tame. Hercules then set the horses free, but they were eventually killed by predators while traversing the slopes of Mount Olympus on their way back to Thrace.

Hercules's ninth labor was very different from any previous. No monster or monstrous animal was involved. What Eurystheus now demanded was that Hercules bring him the belt of Hippolyta, queen of the Amazons. This was no simple task, as the Amazons were a warlike tribe of women and lived on the distant southern coast of the Black Sea. Hippolyta's belt had been a gift from her father, Ares, the god of war, and it was a source of great power. Concerned that conflict with the Amazons would be sure to ensue, Hercules traveled with a group of Greece's finest men. Quite contrary to expectation, Hippolyta gave her belt freely when the request was made. That this task was so easy, however, made the goddess Hera mad, and she disguised herself as Hippolyta's sister, spreading a rumor of the queen's kidnapping by the Greeks. Enraged, the Amazons attacked the men, and in the battle that erupted, Hippolyta was slain. It was Hercules who killed her, believing that somehow he had been betrayed.

Hercules took a roundabout route through Libya from Greece to Erythrea (in what is now Spain) to complete the tenth of his labors. Erythrea was ruled at the time by Geryon, the son of Chrysaor, who happened to be the winged horse Pegasus's brother. Like his father, Geryon was a monster, having three heads and three torsos. He also had prodigious strength. It was Geryon's famed herd of cattle that Hercules was to drive—without their owner knowing—to Greece. Hercules almost got out of Erythrea without incident, though he was spotted before he could drive the cattle to a safe distance. Geryon pursued Hercules, but the hero once again prevailed, fatally shooting the king with his bow and arrow.

His eleventh labor took Hercules back to Libya, this time to what was believed to be the ends or "bottom" of the Earth. Hercules was to pick some of the golden apples growing in a grove tended by the Hesperides, nymphs descended from the Evening Star, Hesperus. The special trees that the Hesperides tended had been a wedding gift to the goddess Hera from Gaia, the great goddess of the Earth, and the trees' golden fruit could guarantee immortality. A sleepless, hundred-headed dragon called Ladon was stationed in the grove to keep guard. By some accounts, Hercules slew the dragon, whom Hera eventually raised to the stars as Draco ("dragon" in Greek), and freely picked some unguarded apples. Alternatively, Hercules did not kill but merely tricked the dragon and the Hesperides. It so happened that the Titan god Atlas stood nearby, supporting the heavens on his shoulders. Atlas, being the Hesperides' neighbor, would be more likely to be able to get the apples, or so Hercules believed. When Hercules offered to take the heavens on his own shoulders for a short while in return for the apples, a weary Atlas readily agreed. Atlas accomplished the task of retrieving the apples but was so relieved to be free of his heavy burden that he refused to take the heavens back. He finally did, however, but Hercules had to trick him, too: he asked Atlas to hold the heavens just for a moment while he put a cushion on his neck.

Hercules's twelfth and final labor was also the most dangerous by far. He was to enter the world of the dead and return with Cerberus, the fearsome hound that guarded entry and exit from the realm of Hades, god of death, and his queen Persephone. Entry into the Underworld while still alive was relatively simple, but only the bravest would ever dare to try. Exit, on the other hand, was very near impossible, as only one of the guard dog's three heads slept at any given time. Though menacing enough, Cerberus was still a dog and happily snapped up a honey cake laced with a sleeping potion from Hercules's hand. Hercules then gathered the sleeping dog in his arms and carried him to Tiryns. After showing him to Eurystheus, he returned Cerberus to his home in the Underworld unharmed.

His labors at long last completed, Hercules's adventures still continued. Among them was his marriage to Deianeira, princess of the kingdom of Calydon on the north coast of the Gulf of Corinth. Hercules won Deianeira's hand in a wrestling match against the river god Achelous, but his victory would become a loss: the marriage led to his ultimate undoing. Hercules was unable to stay in Calydon with his new bride. An accidental killing, caused by drunkenness on Hercules's part, was the reason for his exile. Hercules and Deianeira then left Calydon and soon came to the river Evenus. Swollen with snowmelt, the river raged, dangerous to cross. Hercules felt confident he could make the crossing, but he was concerned about Deianeira. So, Hercules readily accepted when the centaur Nessus offered to carry her across, a fateful decision on his part. As soon as the lustful Nessus got

NORTHERN CELESTIAL QUADRANT III

Deianeira settled on his back, he tried to make off with her. Hercules managed to shoot the centaur and rescue his bride, but not before Nessus gave her a terrible gift. This gift was a small sample of his blood, which Nessus claimed was a love charm that she could use if ever Hercules became interested in another woman. By some accounts, it was Nessus and not Chiron who became Centaurus, the centaur constellation, upon his death.

With the passage of time, there was in fact another woman in Hercules's life. When Hercules brought home the daughter of a defeated king as his concubine, Deianeira dug out Nessus's counterfeit charm. As instructed, she dipped Hercules's cloak in it, having no idea that Nessus's blood was no love potion but, instead, a lethal poison, for the centaur's blood had become tainted by Hercules's arrow—an arrow dipped in the Hydra's blood. No sooner did the cloak touch Hercules's skin than his flesh began to burn, falling away from his body. The pain was unbearable, and there was no antidote. Hercules asked his closest friends to end his life, but each refused this terrible, if merciful, favor. At last, one man, Philoctetes, came forward and lit a funeral pyre for Hercules to ascend as a means to end his own life. As a reward for this humane act, Philoctetes was gifted Hercules's bow and arrow. Hercules, meanwhile, was made a god by the will of Zeus, now joining the company of Hera, who had finally forgiven him. He also had a new bride, Hera's daughter Hebe, and as a deity he could now be seen among the stars. In the constellation named after him, he is captured mid-exploit: he holds his club in his raised right hand and a branch from the Hesperides' tree in his left while lunging forward, pressing one foot down on Ladon (the constellation Draco), poised to strike.

The figure of Hercules is located largely in Northern Quadrant 3, but his left hand, along with the precious Hesperidean branch, extends into Northern Quadrant 4.

Best visibility: June to August (90°N–38°S)

See also the constellations Argo Navis, Aries, Cancer, Centaurus, Draco, Leo, Lyra, Gemini, Hydra, Perseus, Sagitta, and Taurus.

CLASSICAL MYTHOLOGY OF THE CONSTELLATIONS

URSA MINOR

The Lesser Bear

Recognizing her child, Callisto, now a she-bear, froze in her tracks as he approached; but Arcas, frightened at her piercing, steadfast gaze, shrank back and was poised to use his spear to pierce his mother's breast. Luckily Zeus, the omnipotent god, stayed his hand in time, preventing this unspeakable crime. Then, sweeping them both up in a whirlwind through air, he carried them up to the heavens where he placed them as neighboring constellations.

Ovid, *Metamorphoses* 2.500–507

MAIN CHARACTERS:
    Arcas, son of Callisto by Zeus
    Artemis, goddess of the hunt and wild animals
    Callisto, a princess of Arcadia raped by Zeus and transformed into a bear
    Cronus, a Titan god and father of Zeus
    Demeter, goddess of grain and the harvest
    Hades, god of the Underworld
    Hera, queen of the gods and wife of Zeus
    Hermes, the messenger god
    Hestia, goddess of the hearth
    Lycaon, brutal king of Arcadia and father of Callisto
    Poseidon, god of the sea
    Rhea, a Titan goddess and mother of Zeus
    Zeus, king of the gods

There are two very different stories associated with the constellation Ursa Minor (*the Lesser Bear*). The first of these is the better-known myth of Callisto and her son Arcas. Callisto was an extraordinarily beautiful young woman who was a devoted follower of Artemis, goddess of the wilderness and wild animals. Like Artemis herself, Callisto was committed to a life without a husband and marriage. Unhappily for her, Zeus became enamored of Callisto, and in order to get close to her, he changed his own appearance to resemble that of Artemis. This

NORTHERN CELESTIAL QUADRANT III
Ursa Minor

CLASSICAL MYTHOLOGY OF THE CONSTELLATIONS

trick allowed Zeus to force himself on Callisto, who became pregnant from this frightening encounter. In the course of time, she gave birth to a son called Arcas.

An angry, jealous Hera, dismayed at her husband Zeus's infidelity, now changed Callisto into a bear, while the infant Arcas was carried by Hermes to the court of his grandfather, King Lycaon of Arcadia, a legendarily bestial man. Revealing his true, savage nature, Lycaon killed young Arcas and served him to Zeus, chopped up and cooked in a stew. On that particular day, Zeus was disguised as a humble traveler so that he could wander freely among humans to observe firsthand who had fallen into sin. Reputedly, the worst offender was Lycaon. Zeus did not need to taste the stew to know what he was being served and reassembled Arcas, his own son, bringing him back to life. Lycaon, on the other hand, was transformed by the god into a wolf, a punishment memorialized by the constellation Lupus (*the Wolf*).

When Arcas later came upon his mother, Callisto, while he was hunting in the forest, he nearly killed her, never suspecting her true identity. To him, she looked like any other bear. Just in time, however, Zeus intervened and transferred mother and son to the heavens as constellations, both of them in the shape of bears. Callisto is the larger of the two, and Arcas is the smaller, long-tailed bear.

An alternative tale connects Ursa Minor with Zeus's birth and early childhood. Zeus's father, the god Cronus, had learned from a prophecy that one of his children would overthrow him as ruler of all the then-existing gods. Cronus, wanting to avert this fate at all costs, swallowed each of his newborn children in succession as his wife, Rhea, gave birth to them. Cronus had already swallowed five—Demeter, Hestia, Hera, Hades, and Poseidon—by the time Rhea was able to devise a plan to prevent him from continuing. As soon as her sixth child, Zeus, was born, she quickly wrapped a large stone in swaddling clothes and offered it to Cronus, who swallowed it without a second thought. Before Cronus could discover her trickery, Rhea sent the baby Zeus away to Mount Ida on the island of Crete, where he was lovingly raised by nymphs. It was two of his Cretan nurses that Zeus later raised to the night sky in thanks, curiously giving them the shape of starry bears: the Lesser and the Greater Bear. As for his unfortunate siblings, Zeus saw to it that Cronus regurgitated them.

Ursa Minor is notable for containing the North Star, Polaris, which is the brightest star in the constellation and has traditionally been important for navigation. Polaris is located at the tip of the bear's long tail. Ursa Minor's seven main stars, which include Polaris, constitute the asterism (small group of stars) known today in North America as the Little Dipper.

Best visibility: May to June (90°N–0)

See also the constellations Lupus and Ursa Major.

NORTH CELE QUAD

HERN
STIAL
RANT
IV

NORTHERN CELESTIAL QUADRANT IV
Aquila

NORTHERN CELESTIAL QUADRANT IV

AQUILA

The Eagle

Zeus made the eagle king of all birds, appointing it the guardian of his sacred scepter as well as of his throne.

Antoninus Liberalis, *Tales of Metamorphosis 6*

MAIN CHARACTERS:

Apollo, the Greek god of prophecy, music, and medicine
Cronus, a Titan god and the father of Zeus
Cyclopes, one-eyed giants who made Zeus's thunderbolts
Ganymede, a Trojan prince and cupbearer of Zeus
Helios, god of the Sun
Hercules, the greatest of all Greek heroes and known for his Twelve Labors
Periphas, an early king of the territory of Athens
Prometheus, a Titan god who helped humankind
Zeus, king of the gods

The constellation Aquila (*the Eagle*) represents the god Zeus's sacred bird. Authors in antiquity offered a number of explanations for Zeus's choice of this particular bird, which symbolized his rule and became his avian attendant. By some accounts, it was because eagles were believed to be the only birds that fly directly toward the rising sun, this being evidence of their inherent royalty and divinity. The Sun was Helios, a powerful god in his own right. Some authors, on the other hand, noted that the eagle was the only bird that could easily carry Zeus's weapons, thunderbolts forged by the one-eyed giants called Cyclopes. Further, it was this bird that appeared over the altar (Ara) at which Zeus was making sacrifice in the hope of securing victory over the older-generation Titan gods, among whom was his father, Cronus. On this occasion, the eagle's appearance in the sky was unambiguously a good omen, since Zeus and his siblings were victorious, leaving them to rule the heavens, the Earth, and the Underworld.

    There was an alternate tradition that the eagle of Zeus was originally a wealthy, pious early king of Attica, the territory of Athens. He was so devout and generous that he came to be regarded as a living god, even addressed by some as Zeus himself. That a mortal should receive the same respect and reverence as he did angered Zeus, and he resolved to destroy this man, Periphas, and his entire family. Had Apollo not intervened, begging Zeus that he be spared, Periphas would have lost everything to Zeus's fiery thunderbolts. Instead of killing his rival, Zeus transformed him into an eagle, and Apollo stationed him at the side of Zeus's throne.

While in Zeus's service, the eagle executed a range of tasks. Among them was seizing the Trojan prince Ganymede from the slopes of Mount Ida where he was tending his sheep (though some said it was Zeus himself, disguised as the eagle, that did this). The handsome young Ganymede would become Zeus's cupbearer (Aquarius), ensuring that all the gods' goblets were filled while they were feasting. Another notable task that the eagle was given—a gruesome one!—was to eat the liver of Prometheus, a Titan god famous for helping humankind but disliked by Zeus for helping mortals too much. It was Prometheus who first brought humans fire, and it was he who tricked the gods into accepting fat-wrapped bones as their typical offering, not succulent meat, whenever animal sacrifices occurred. As punishment, Zeus caused Prometheus to be chained to a massive rock on the Caucasus Mountains and his liver to be eaten away by the eagle. The punishment was meant to be eternal, and it would have been—because the liver continuously regenerated as it was eaten—had Hercules not happened upon poor Prometheus and killed the eagle with one of his lethal arrows. Both the eagle's snatching of Ganymede and its service as Prometheus's torturer earned it a place among the stars, where it is commemorated alongside Hercules's arrow, the nearby constellation Sagitta. Wings outstretched, it flies along the Celestial Equator with Zeus's thunderbolts in its talons, its legs extending into Southern Quadrant 4.

Best visibility: July to August (78°N–71°S)

See also the constellations Aquarius, Ara, Hercules, and Sagitta.

NORTHERN CELESTIAL QUADRANT IV

CEPHEUS

Cepheus

At the monster's death, the shores echo with cries of joy and applause, even reaching the palaces of the gods above. Cassiopeia and Cepheus, the rescued maiden's father, rejoice and proclaim their new son-in-law the deliverer and savior of their house. Now, her chains undone, here comes the maiden—she was the prize and the cause of this heroic task.

Ovid, *Metamorphoses* 4.735–739

MAIN CHARACTERS:
- Ammon, an Egyptian god identified by the Greeks with Zeus
- Andromeda, an Ethiopian princess rescued by Perseus from the sea monster Cetus
- Athena, the goddess of wisdom and defensive war
- Cassiopeia, the queen of Ethiopia and mother of Andromeda
- Cepheus, the king of Ethiopia and father of Andromeda
- Cetus, the sea monster sent by Poseidon to ravage Ethiopia's coast
- Medusa, the snake-haired Gorgon beheaded by Perseus
- Poseidon, god of the sea

Cepheus was a king of Ethiopia and blessed with a lovely daughter, Andromeda, and a beautiful wife, Cassiopeia. All was well in his kingdom until Cassiopeia stupidly bragged that she herself, or by some accounts her daughter, was more beautiful than even the Nereids, fair nymphs of the sea.

For this prideful boast, Ethiopia and Cepheus were punished. The god Poseidon, sympathizing with the Nereids, sent a flood upon the land and then a sea monster to plague Ethiopia's populace. Not knowing what else to do, Cepheus made a journey to Libya, site of the sacred oracle of Ammon. There he inquired what remedy there might be for both the plagues upon his lands. The response he received was a most unwelcome one: he should sacrifice his own daughter to Poseidon. More specifically, he was to chain Andromeda to a sheer rock face on the coast as fodder for Cetus, the sea monster of Poseidon. Cepheus wrestled with this

NORTHERN CELESTIAL QUADRANT IV
Cepheus

notion. Sacrificing his daughter was just too much for him to bear, but the people of his kingdom pressured him to comply with the oracle's instructions.

And so Andromeda was chained up for the sea monster to take her away, but the hero Perseus happened to pass overhead, flying through the air on his winged sandals and still holding the severed head of Medusa. He fell instantly in love with this maiden, who was in clear distress, and offered his assistance to her parents in exchange for her hand in marriage. Cepheus and Cassiopeia readily agreed, forgetting in that moment that Andromeda had already been promised to Cepheus's brother. In any event, Perseus slew the monster, rescued Andromeda, and successfully fought (or outwitted) Cepheus's brother. By some accounts, Cepheus then attempted to dissuade Andromeda from leaving with Perseus, but with no success. Perseus and Andromeda were married and had seven children together. When the goddess Athena placed Perseus and Andromeda among the stars upon their deaths, Poseidon did the same for Cepheus, Cassiopeia, and Cetus the sea monster. In this way, since all these constellations lie in close proximity to each other, this family's story continues to play out eternally.

Best visibility: September to October (90°N–1°S)

See also the constellations Andromeda, Cassiopeia, Cetus, and Perseus.

CYGNUS

The Swan

> Remember how often Zeus changed his form to some lesser creature, and he's the mighty god who controls the weather and the clouds: as a bird he fluttered his white wings and sang more sweetly than a dying swan, and later as a wild bull he knelt down tamely, lowering his back so that maidens could play with him.

Seneca, *Phaedra* 299–304

MAIN CHARACTERS:

Agamemnon, a king of Mycenae and commander of the Greeks who sailed to Troy

Callisto, a young woman pursued by Zeus and transformed into a bear

Europa, the Tyrian princess kidnapped by Zeus disguised as a tame white bull

Helen, a queen of Sparta and the cause of the Trojan War
Hera, queen of the gods
Io, the daughter of a river god who was seduced by Zeus and transformed into a cow
Leda, mother of Helen of Troy and of the twins Castor and Pollux
Menelaus, a king of Sparta and husband of Helen of Troy
Paris, the Trojan prince who made off with Helen
Tyndareus, a king of Sparta and husband of Leda
Zeus, king of the gods

The constellation Cygnus (*the Swan*) is a representation of Zeus disguised as a swan in flight, coursing along the Milky Way. Zeus, mightiest of the Greek gods, was married to the goddess Hera but, nonetheless, was famous—or infamous—for his roving eye. When seducing mortal women, it was usual for him to transform himself, often as a lure but sometimes to evade detection by his wife. He approached the Phoenician princess Europa as a lovely, tame bull, and incidentally, that bull, too, can be seen among the stars as the constellation Taurus. When Zeus made advances upon the princess Io, he changed himself into a cloud. This he did to hide himself and Io from Hera. Zeus approached the chaste maiden Callisto disguised as the goddess Artemis. Poor Callisto was later transformed into a bear by an angry Hera, though Zeus did ultimately raise her to the heavens as the constellation Ursa Major (*the Greater Bear*). In order to seduce Leda, queen of Sparta, Zeus became a swan, and in the course of this encounter fathered a daughter, the lovely Helen, and Pollux, a noble son.

As it happened, Leda was impregnated by her husband, King Tyndareus, on the same day, and Tyndareus became father to Helen's evil sister, Clytemnestra, and to Castor, Pollux's mortal twin. The twin boys, Castor and Pollux, would be immortalized in the constellation Gemini. Helen, on the other hand, would become the cause of the famous ten-year Trojan War. While Helen was married to Menelaus, who had become king of Sparta, the Trojan prince Paris came to visit. When Paris finally returned home, he did so with Helen, taking her away in the dead of night. It was an unspeakable affront to Menelaus, his generous host. To get her back, Menelaus and his brother Agamemnon, king of powerful Mycenae, assembled a fleet of 1,000 ships to sail from Greece to Troy. The battle was unrelenting and lasted a full decade until—through the stratagem of the Trojan Horse, a false, warrior-filled peace offering to the Trojans—the city fell to the invading Greeks.

Best visibility: August to September (90°N–28°S)

See also the constellations Gemini, Taurus, and Ursa Major.

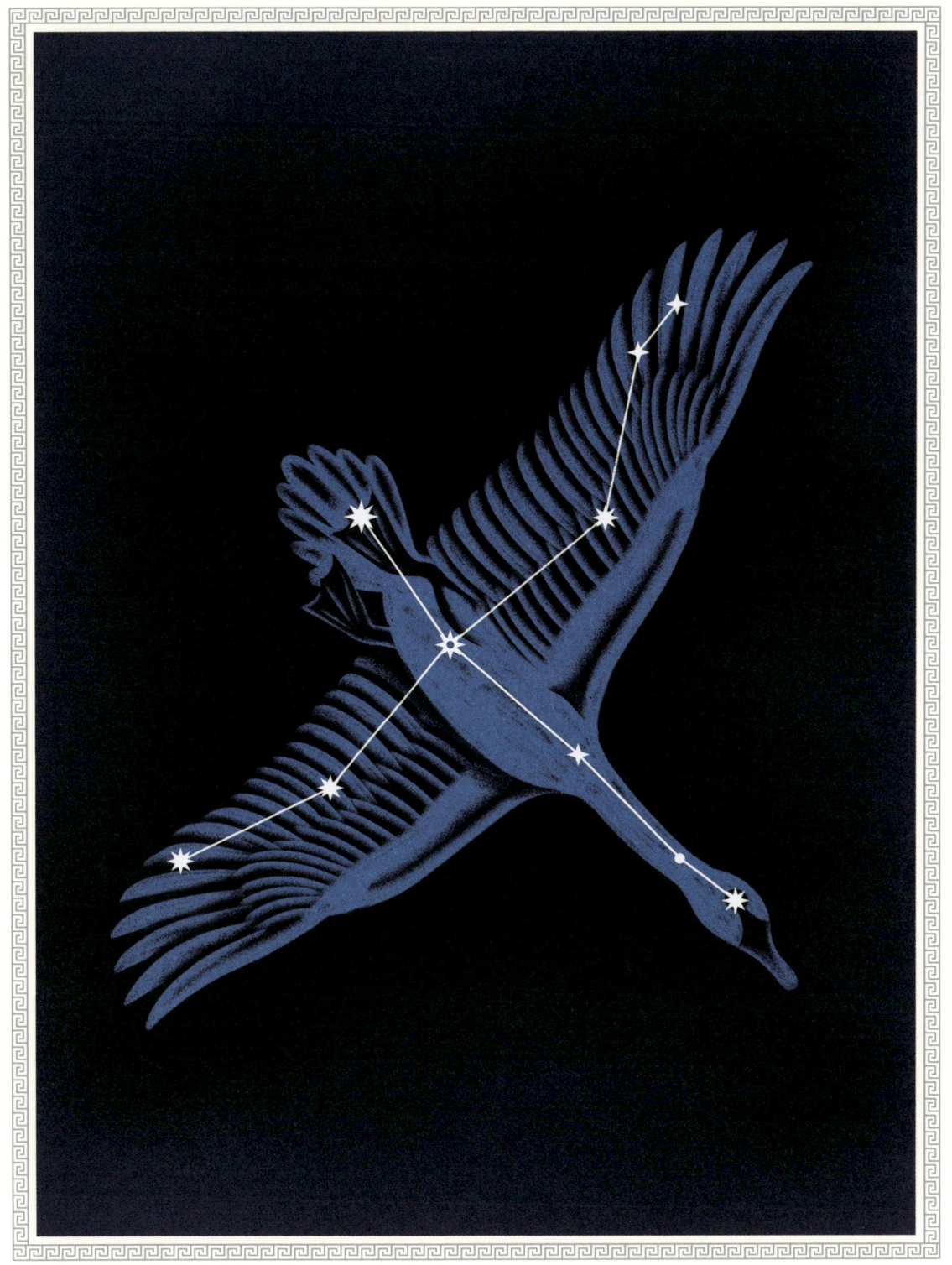

NORTHERN CELESTIAL QUADRANT IV
Cygnus

CLASSICAL MYTHOLOGY OF THE CONSTELLATIONS

DELPHINUS

The Dolphin

> The pirate Medon's body began to blacken and his spine arched, curving to extreme. "What strange creature are you turning into?" cried his comrade Lycabas, but, even as he spoke, his mouth widened in a kind of grin, his nose curved out, his skin turned hard and scaly. Their friend Libys, trying to pull back the oars, saw his hands suddenly shrink—they weren't hands any longer—instead they should be called fins! Another man, when trying to grab the towing rope, discovered that he had no arms and jumped, limbless, bending backwards into the sea. His tail forked, curved like a sickle moon. Now they leaped around the ship, showering its hull with an abundance of ocean spray.
>
> Ovid, *Metamorphoses* 3.671–684

MAIN CHARACTERS:
   Amphitrite, a sea goddess loved by Poseidon
   Arion, a legendary singer and musician
   Dionysus, the god of wine, vegetation, and the theater
   Poseidon, god of the sea
   Semele, a princess of Thebes and mother of Dionysus
   Zeus, king of the gods and father of Dionysus

There are three very different stories that explain the origins of the constellation Delphinus (*the Dolphin*). The best-known of these is the tale of the god Dionysus and his kidnapping by a most unwise group of pirates. Dionysus, who is chiefly known as the god of wine, was also a god of vegetation and a shapeshifter, able to take on the appearance of humans and animals alike.

Dionysus was a son of Zeus and Semele, a princess of the Greek city of Thebes. When Semele became pregnant by Zeus, she divulged this fact to her sisters. Out of jealousy that she had captured the attentions of a god, Semele's sisters urged her to ask Zeus to prove that he was actually a god and not an imposter. This Semele did, and Zeus revealed himself to her in his full divinity—a dangerous thing to do. Gods usually appeared to humans in a toned-down form. Encountering Zeus

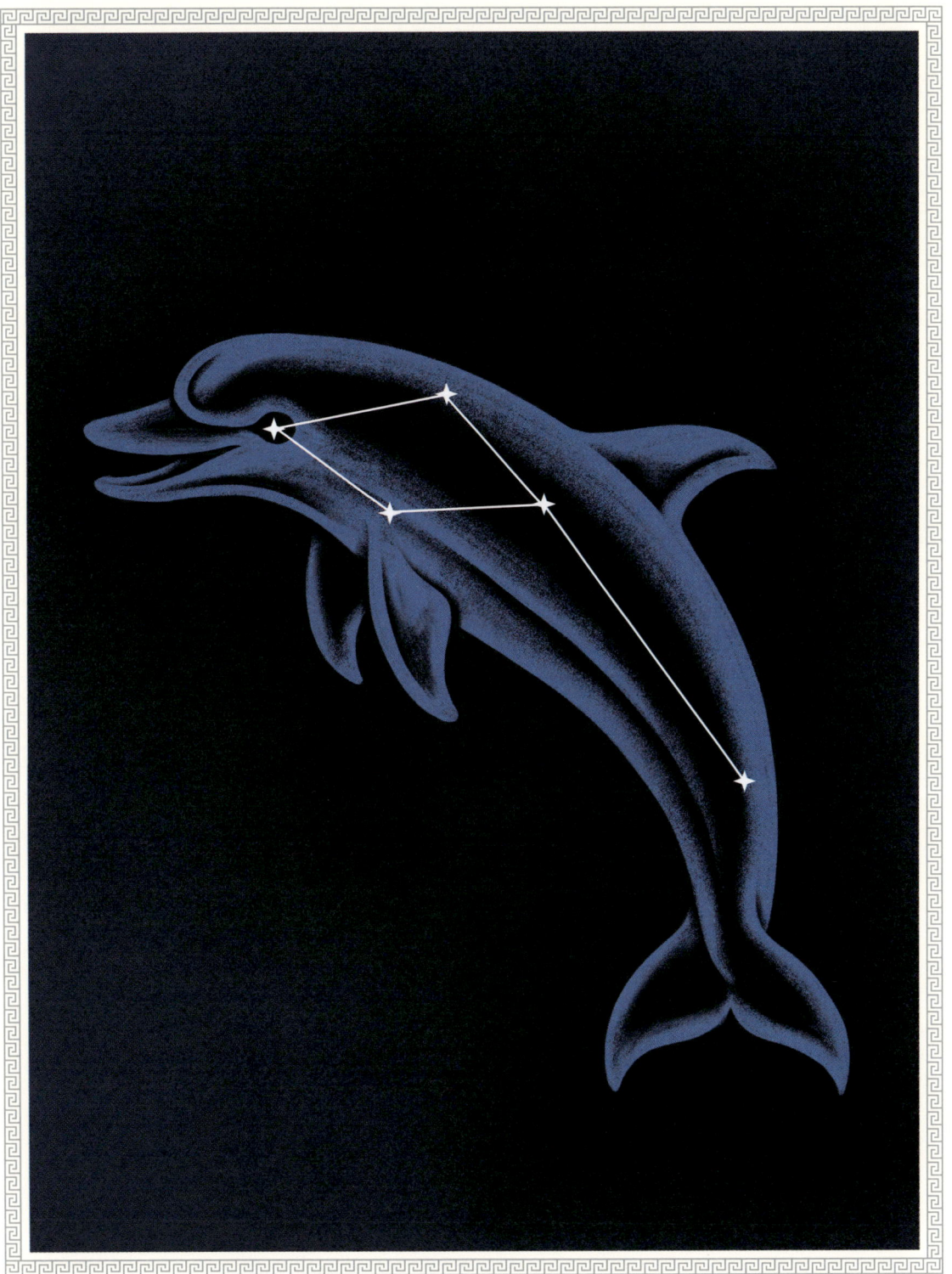

NORTHERN CELESTIAL QUADRANT IV
Delphinus

in all his heavenly splendor was too much for Semele, who burst into flames in his presence, and the god was unable to save her. He did, however, save the child Dionysus, placing the fetus in his thigh until he was ready to be born.

Upon his birth, Dionysus was sent to the mountains of Nysa (somewhere in Thrace, Asia, or Africa) to be raised by the resident nymphs, but as he matured, he decided to make his way to Mount Olympus, the home of the other gods. On his way, he asked a group of pirates to give him passage to the island of Naxos on their ship, and they agreed, but avarice would get the better of them. To them, Dionysus looked like the son of a wealthy nobleman and likely to command a substantial ransom if they kidnapped him. So that is what they decided to do, and they sailed right by Naxos. Dionysus, of course, insisted that they turn around, but the pirates simply mocked him. Who was this youth to tell them what to do?

Greed and stupidity were their undoing. Suddenly the music of a flute could be heard, an eerie sound, and ivy grew to cover the ship's oars and masts, stopping it from moving. The youth himself was decked in grapevine, and phantom tigers, lynx, and panthers circled around him. In abject terror, the pirates tried to flee and leaped into the water. But as they leaped, the god transformed them into dolphins. According to this story, it is because they were once humans that dolphins are drawn to people and follow along with ships, and to memorialize this event, Dionysus placed a dolphin, Delphinus, in the heavens. Dionysus himself did not appear among the stars, but the crown he later presented to his bride became the constellation Corona Borealis (*the Northern Crown*).

A second myth centers on the sea god Poseidon, who fell in love with the sea goddess Amphitrite and wished to marry her. Amphitrite wanted nothing to do with this and hid so successfully that Poseidon could not find her. The god then enlisted a dolphin to help him in his search. Not only did the dolphin find her, but he also persuaded her to accept Poseidon's attentions. As a reward, Poseidon placed him among the stars.

A story about the legendary musician and singer Arion presents another explanation for the dolphin's presence in the heavens. Arion, it was said, had amassed quite a fortune while performing on a tour of Sicily, a fact of which the crew of the ship carrying him back home to the city of Corinth was aware. They decided that the most expedient way to get their hands on Arion's money was simply to throw him overboard. A payment for his life, which Arion offered, had no appeal for them when they could have all the riches that Arion had. So Arion asked for one last favor: could he sing just one final song? This favor was granted. Much to the men's surprise, a pod of dolphins arrived, cavorting around the ship, when Arion began to sing. More surprising still, Arion jumped overboard onto a dolphin's back, and it carried him home to safety. As a memorial to the dolphin's good deed, the gods lifted it to the heavens.

Best visibility: August to September (90°N–69°S)

See also the constellation Corona Borealis.

EQUULEUS

The Foal

I sing of you, too, god of the sea, for whom the Earth poured forth the neighing horse when struck by your mighty trident.

———

Vergil, *Georgics* 1.12–14

MAIN CHARACTERS:
  Athena, goddess of wisdom and defensive war
  Bellerophon, the hero who tamed Pegasus and slew the Chimera
  Castor, a brother of Helen of Troy and twin of Pollux
  Medusa, a snake-haired Gorgon and mother of Pegasus
  Pegasus, the winged horse
  Pollux, twin brother of Castor and one of the famous twins called the Gemini
  Poseidon, god of the sea and of earthquakes

Equuleus (*the Foal*) became associated with a number classical antiquity's fabled horses, though there is no detailed narrative from Greek or Roman authors explaining how the foal became a constellation. Since this constellation is directly adjacent to the constellation Pegasus, it is not surprising that Equuleus, a smaller and fainter constellation, was thought of by some as Pegasus's foal. The famous winged horse Pegasus was a child of the snake-haired Gorgon Medusa and—a gruesome miracle—was born from his mother's severed neck when the hero Perseus beheaded her. Pegasus is best known as a favorite of the Muses, with whom he kept company on Mount Helicon, and, later, as the horse of the hero Bellerophon, who tamed him and would later slay the hybrid monster—part lion, dragon, and goat—known as the Chimera.

Equuleus was also seen as the horse created by the god Poseidon in his famous contest with Athena. Both gods were eager to become the patron deity of Athens,

NORTHERN CELESTIAL QUADRANT IV

Equuleus

and a contest between them seemed the fairest way to settle this. In her bid for the judges' favor, Athena produced an olive tree, which is replicated by an olive tree now growing on the Acropolis. Poseidon, in turn, struck the earth with his trident, and from the fissure left by his weapon sprang the first horse, his gift to the city. After careful deliberation, the judges decided in favor of Athena, believing that hers was the more valuable gift. Olive oil did, in fact, become the mainstay of the Athenian economy, but horses, the most highly valued of all domesticated animals in ancient Greece, were enormously important, too.

In an alternate tradition, Equuleus was instead the swift horse that belonged to Castor or Pollux, the heroic twins memorialized by the constellation Gemini.

Whatever the reason for Equuleus's presence in the sky, it is only the foal's head that appears.

Best visibility: September (90°N–77°S)

See also the constellations Gemini and Pegasus.

LYRA

The Lyre

The Bacchantes, in a frenzy, set upon and killed the poet Orpheus, then dismembering his lifeless body. His scattered limbs were gathered by the Muses, who gave them burial, and as the greatest favor they could confer, they made his lyre an eternal memorial, pictured with stars, among the constellations. This was the will of Apollo and of Zeus.

Hyginus, *Astronomica* 2.7.1

MAIN CHARACTERS:
  Apollo, Greek god of music and father of Orpheus
  Aristaeus, a minor agricultural deity
  Bacchantes, female celebrants of the god Dionysus
  Cerberus, the triple-headed guard dog of the Underworld
  Charon, ferryman of the Underworld
  Dionysus, the egalitarian Greek god of wine
  Eurydice, Orpheus's wife

Hades, god of the Underworld
Hercules, the greatest of all Greek heroes and famed for his Twelve Labors
Hermes, the messenger god
Jason, the hero who spearheaded the launch of the ship *Argo*
Muses, patron goddesses of the arts
Orpheus, the most skilled of all Greek poets and musicians
Persephone, queen of the Underworld
Zeus, king of the gods

The small but bright constellation Lyra (*the Lyre*) is typically associated with the tragic tale of Orpheus, a legendarily exquisite singer, poet, and musician. His instrument of choice was the lyre, a stringed instrument resembling a small harp but with a sounding box. The lyre was said to have been invented by the messenger-god Hermes on the very day of his birth. A precocious and intelligent child, Hermes grabbed a tortoise, separated the creature from its shell, and covered the shell's cavity with ox hide. The shell became the instrument's sounding box, and to this a stringed yoke of horns was fastened.

Hermes gifted the lyre to his brother Apollo, god of music, medicine, and prophecy. Apollo, in turn, gave the lyre to Orpheus, who was by some accounts his son, and the Muses, patron goddesses of the arts, taught Orpheus how to play it. Orpheus became so skilled a musician that he was even able to charm animals, trees, and stones with his playing. Mesmerized by his tunes, these would follow after him.

In the course of his life, Orpheus traveled widely and had a host of adventures, among them the fabled expedition of the ship *Argo* (Argo Navis). *Argo* was the first seafaring ship ever built, and it was custom-built for the hero Jason, who with a group of valiant men sailed across the Aegean to the far shores of the Black Sea. What they sought on those distant, foreign shores was a famous golden fleece, now the precious keepsake of the realm of Colchis. The men who accompanied Jason were called the Argonauts, and Orpheus was in their number.

On the conclusion of this adventure, Orpheus fell in love with and married a nymph named Eurydice. This great love marked the start of Orpheus's troubles. One fateful day, Eurydice caught the eye of Aristaeus, a minor deity associated with agriculture and beekeeping. Aristaeus ran after her—his intent was rape—and Eurydice fled, not watching where she stepped. As misfortune would have it, a venomous snake was in her path, and she would perish of its bite. A heartbroken Orpheus now dared to do what few were brave enough to attempt, namely, to enter the Underworld while still alive. With his music, he charmed Charon, the

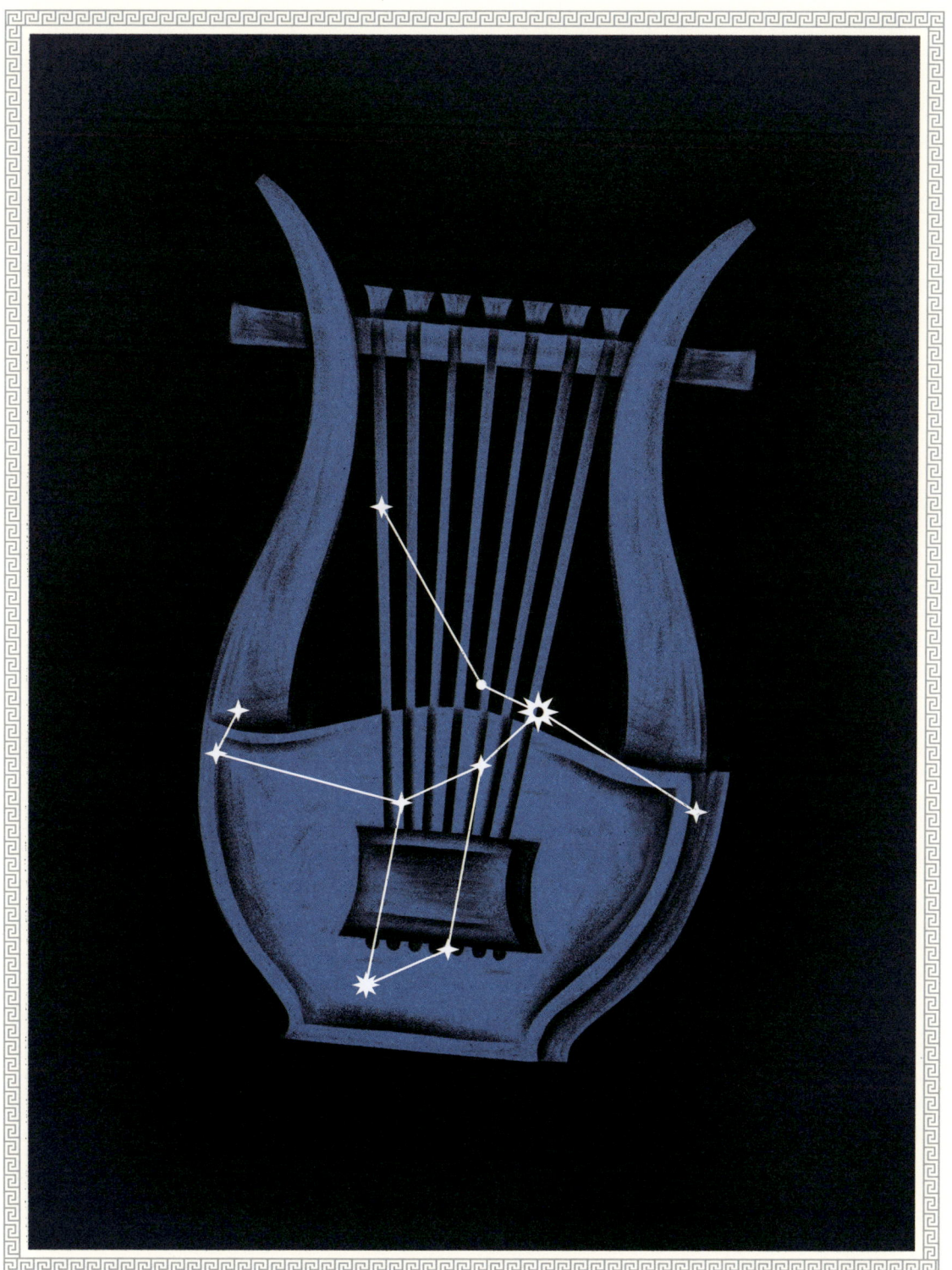

NORTHERN CELESTIAL QUADRANT IV
Lyra

ferryman of the dead, and he charmed the terrifying triple-headed watchdog Cerberus, too. Even Hades, lord of the Underworld, and his wife, Persephone, were swept away by his playing. They were so moved that they granted Orpheus's wish: the return of Eurydice. There was just one condition, this being that Orpheus should not look back as Eurydice followed him to the world of the living. But, as Orpheus and Eurydice made their ascent, he became overcome with worry. Was she still behind him? Had she fallen back? Unable to help himself, he turned, only then to see Eurydice descend into the gloom.

A despondent Orpheus wandered the wilds of Thrace, his homeland in Southeast Europe, and as he wandered, he caught the attention of many women who wished to win his love. Steadfast in his devotion to the memory of Eurydice, Orpheus did not give way. At last he came across a group of Bacchantes, devotees of the god Dionysus. Known primarily as the god of wine, Dionysus was also a great equalizer, blurring distinctions between all people, and his worshippers would literally step out of their normal selves, entering a state of temporary madness. The Bacchantes, too, made advances on Orpheus, and, true to form, he rebuked them. Enraged at this slight, and in a frenzied Bacchic trance, the Bacchantes set upon him and tore him limb from limb. They tossed his still-singing head and his lyre into the Hebrus River, whose currents carried both head and lyre to the island of Lesbos. There, the Muses placed Orpheus's head in a sacred cave and placed the lyre in Apollo's temple. As a further gesture in honor of the great musician, the Muses placed an image of the lyre among the stars, as Apollo and Zeus desired.

An alternate, less widely known myth associated with the constellation Lyra is the story of Hercules and Linus, Orpheus's brother. Linus, too, was a skilled musician and had become the young Hercules's instructor in the lyre. When, in the course of a lesson, Linus criticized Hercules's playing, the latter struck him in a fit of anger, killing him. In this variant, the heavenly lyre commemorates the talented Linus.

Best visibility: July to August (90°N–42°S)

See also the constellations Argo Navis, Aries, and Hercules.

NORTHERN CELESTIAL QUADRANT IV

PEGASUS

Pegasus

While Medusa and her snakes were in the grips of deepest sleep, Perseus sheared her head right off her neck; and from the bloody stump sprang two children: swift-flying Pegasus and his brother Chrysaor.

Ovid, *Metamorphoses* 4.785–787

MAIN CHARACTERS:
Andromeda, the princess rescued by Perseus from a sea monster
Athena, goddess of defensive war and patron goddess of Athens
Bellerophon, a Greek hero tasked with slaying the Chimera
Chrysaor, the brother of Pegasus and a sword-wielding giant
Iobates, a king of Lycia and father of Stheneboea
Medusa, a snake-haired Gorgon slain by Perseus
Nemean Lion, the monstrous lion killed by Hercules as his first labor
Pegasus, a winged horse and child of Medusa
Perseus, the hero who beheaded Medusa
Poseidon, god of the sea
Stheneboea, a queen of Tiryns who wrongfully accused Bellerophon
Zeus, king of the gods

The constellation Pegasus is named after the only winged horse in Greek mythology. Improbably enough, this astonishing creature was a child of the snake-haired Gorgon Medusa and for this reason is linked to the tale of Perseus and Andromeda, both of whom became constellations. However, Pegasus has his own story, too, beginning with his strange conception and birth.

Medusa, Pegasus's mother, had not always been a hideous monster. She was one of three sisters, together known as the Gorgons. Medusa was not only beautiful but also the only sister to be mortal. In other words, her siblings could not be killed. The god Poseidon, lord of the sea, fell in love with her and pursued her, eventually making love to her in a temple of the goddess Athena. This act was a violation of a sacred space, and for this infraction, Athena punished Medusa.

NORTHERN CELESTIAL QUADRANT IV

Pegasus

NORTHERN CELESTIAL QUADRANT IV

Medusa's once lovely, curling locks became writhing snakes and her face so hideous that it turned anyone who looked upon it to stone.

When the hero Perseus was sent on a mission to kill Medusa, he had the assistance of the gods, who equipped him with a cap that made him invisible, winged sandals with which to make a quick escape, and a shield that was as shiny as a mirror. Perseus was able to approach Medusa soundlessly and used the shield to locate her in the Gorgon's lair. Relying only on a reflection of his prey and never having to look directly at her, Perseus swiftly beheaded her with his sword. From her severed neck sprang Pegasus and the giant called Chrysaor. Medusa, as it happened, had become pregnant with these creatures by Poseidon during that fateful encounter in Athena's temple.

Pegasus became a favorite of the Muses, patron goddesses of the arts, and also became integral to the life of the hero Bellerophon and his many death-defying exploits. As a young man, Bellerophon had accidentally killed his brother, and, as was typical in such cases, he was sent away in exile from his homeland. He was received hospitably by the king of Tiryns and even more hospitably by the queen, who fell in love with him. This queen, Stheneboea, made advances on him, but Bellerophon would have none of it. This prompted her to accuse Bellerophon of rape—out of spite and in order to have her vengeance. The enraged king sent Bellerophon away to the kingdom of Stheneboea's father, Iobates, in distant Lycia (southwestern modern Turkey). He secretly sent a letter to Iobates, too, explaining the need for Bellerophon's punishment. The punishment that Iobates devised was a task that Bellerophon was not expected to survive. This task was to hunt and kill the Chimera, a fire-breathing monster with the head and forequarters of a lion, the belly of a goat, and the tail of a serpent. The Chimera was a sibling of a host of other monsters, including the Nemean Lion (Leo), Ladon the dragon (Draco), and the Hydra, all three of whom became constellations.

Bellerophon now knew he needed help, and like many others in need of guidance, he sought the advice of an oracle. Pegasus, he was told, was the key to his success, and the horse could be tamed—his to ride—if he first visited the goddess Athena's nearby temple. While there, Bellerophon fell asleep and dreamt that he had a golden bridle. Upon awaking he found such a bridle in his hands, a gift from Athena, and he easily fitted it to Pegasus, who had lowered his head to drink. Now astride Pegasus's back, Bellerophon was able to ambush the Chimera and slay it with his arrows.

Alarmed that Bellerophon had returned to Lycia unharmed, Iobates set the hero a further series of seemingly impossible tasks, among them fighting the Amazons. In every instance, Bellerophon prevailed and, as a result, he became overconfident in his abilities. He believed, wrongly, that he had earned a place among the gods. Leaping onto Pegasus's back, he directed the horse up into the heavens, but Zeus would have none of this. The god sent a gadfly to sting Pegasus on the rump, and the bucking horse sent Bellerophon falling to the Earth below. While Bellerophon now lived a life of dishonored solitude, Pegasus was stabled with Zeus's own horses and an image of him, showing only half his galloping body, was placed by Zeus among the stars as a constellation. While Pegasus's head and forequarters are located in Northern Quadrant 4, a small portion of his back and flanks extends into Northern Quadrant 1.

Best visibility: September to October (90°N–53°S)

See also the constellations Andromeda, Draco, Hydra, Leo, and Perseus.

SAGITTA

The Arrow

With this arrow Apollo killed the Cyclopes who forged the thunderbolt that caused his son Asclepius's death. Apollo then buried the arrow in the Hyperborean mountains of the farthest North, but when Zeus pardoned Asclepius, making him divine, the arrow was carried by the winds back to Apollo. For this reason it is now among the constellations, or so it is believed.

―――

Hyginus, *Astronomica* 2.15.6

MAIN CHARACTERS:
Apollo, god of prophecy, medicine, and archery
Asclepius, a son of Apollo who became a healing deity and a constellation
Cronus, a Titan god and the father of Zeus
Cyclopes, one-eyed giants and children of the Earth goddess Gaia
Gaia, goddess of the Earth
Hecatoncheires, hundred-handed giants and siblings of the Cyclopes
Hesperides, nymphs that tended Hera's sacred apple grove
Hydra, a many-headed dragon killed by Hercules
Ladon, the dragon that guarded the apples of the Hesperides
Uranus, god of the heavens

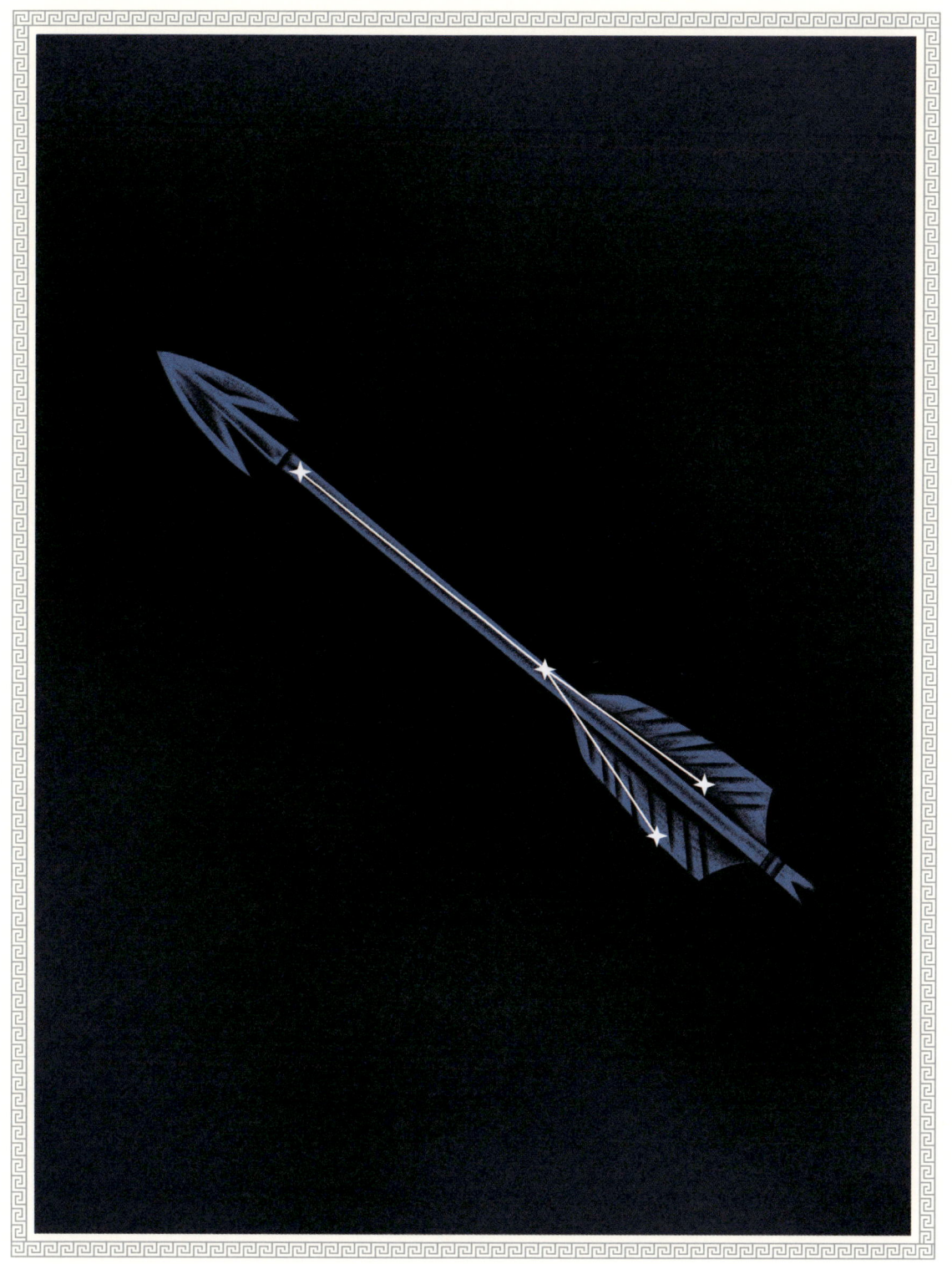

NORTHERN CELESTIAL QUADRANT IV
Sagitta

CLASSICAL MYTHOLOGY OF THE CONSTELLATIONS

Prometheus, a Titan god who was punished by Zeus for always helping humans
Python, a dragon that guarded Delphi and was slain by Apollo
Zeus, king of the Greek gods

The constellation Sagitta (*the Arrow*) is associated with several myths, all of them centering on classical mythology's most renowned archers. Among mortals, the most renowned archer was the hero Hercules, whose arrows were both unfailingly accurate in their aim and absolutely lethal, as their tips had been dipped in the dreaded Hydra's poisonous blood. Among gods, it was Apollo, god of prophecy, medicine, music, and also of archery.

Although Apollo used his bow and arrows on many occasions, among them slaying the dragon Python that guarded the site of Apollo's future oracle at Delphi, it was his assault on the one-eyed Cyclopes that garnered his arrows particular interest. These Cyclopes were children of Gaia, goddess of the Earth, and her consort Uranus, god of the heavens, and were unrelated to the Cyclops that Odysseus encountered on his arduous ten-year return from Troy. When the Cyclopes and their siblings the Hecatoncheires ("Hundred-handers") were born, their father, Uranus, detested these monsters at first sight and shoved them back inside their mother, causing her enormous pain. In agony, Gaia approached all her other children asking for help, but none dared to incur the wrath of their father, apart from the youngest, Cronus. Cronus lay in wait for his father and castrated him when he came to lie with Gaia that night. This left Cronus, temporarily, king of the gods. Later, Cronus was defeated and replaced as king of the gods by his own son Zeus, who released the Cyclopes from their prison in Mother Earth. In gratitude, the Cyclopes forged thunderbolts for Zeus to use as his signature weapon. Zeus's victims included Asclepius, a son of the god Apollo who had become such an expert healer that he achieved what had previously been impossible: bringing the dead back to life. Since immortality is what primarily distinguished gods from mortals, Zeus perceived Asclepius as a threat and killed him with a thunderbolt. A grief-stricken Apollo knew that he could never take out his anger on Zeus directly, so he drew his bow to kill the Cyclopes instead, striking them all with a single arrow. For this Zeus punished him by making him the servant of Admetus, a mortal man, for the period of one year. As for Asclepius, since he was the son of a god, Zeus later placed him among the stars as the constellation Ophiuchus (*the Serpent-Holder*). At the same time, Asclepius became divine, and Apollo's arrow joined him among the stars.

The arrows of Hercules, on the other hand, featured in a number of his famous Twelve Labors, although two particular uses of them were especially noteworthy.

One was Hercules's elimination of a large flock of crane-like birds that had settled around Lake Stymphalus in the central Peloponnese. Unusually, and terrifyingly, these birds shot their metallic bronze feathers like arrows and were terrorizing the region's inhabitants. A skilled archer, Hercules killed many of these birds with his own arrows after startling them from their roosts with a rattle. Those that remained alive were frightened away. When Hercules later went to fetch the famed Golden Apples of the Hesperides, which were guarded by the sleepless dragon Ladon (Draco), he came across the Titan god Prometheus, who had been chained to an outcropping of rock in the Caucasus Mountains. Prometheus was being punished by Zeus for helping humans too much—at the expense of the gods, in Zeus's opinion—and for this reason he was not only tied in chains but also pecked at by Zeus's eagle (Aquila). This eagle was busy eating Prometheus's liver, which grew back as quickly as it was eaten away. Taking pity on Prometheus, Hercules shot the eagle and unfastened Prometheus's shackles. One of his arrows was then placed in the heavens by the gods near the constellations Aquila the eagle to the south and Cygnus the swan to the north. Like the swan and eagle, the constellation Sagitta's arrow soars through the heavens along the Milky Way.

  Best visibility: August (90°N–69°S)

  See also the constellations Cygnus, Draco, Hercules, Hydra, and Ophiuchus.

SOUT
CELE
QUAD

HERN
STIAL
RANT
I

SOUTHERN CELESTIAL QUADRANT I
Cetus

SOUTHERN CELESTIAL QUADRANT I

CETUS

The Sea Monster

Suddenly, a thunderous sound echoed from the deep, and there, advancing over the broad expanse of ocean, a monstrous creature loomed up, its chest rearing over the wide waves. Seeing it, the maiden Andromeda screamed.

Ovid, *Metamorphoses* 4.688–691

MAIN CHARACTERS:
Andromeda, princess of Ethiopia
Athena, Greek goddess of wisdom and defensive war
Cassiopeia, queen of Ethiopia and mother of Andromeda
Cepheus, king of Ethiopia and father of Andromeda
Cetus, a sea monster
Medusa, one of the three Gorgon sisters
Perseus, Andromeda's rescuer and Medusa's slayer
Poseidon, Greek god of the sea

The sea monster Cetus played a pivotal role in the lives of the Greek hero Perseus and the Ethiopian princess Andromeda. The moment that Perseus laid eyes on Andromeda, it was love at first sight, but in order to make her his bride, he would have to rescue her from a cruel, certain fate. She had been shackled to a towering face of rock as an offering to Cetus, the sea monster sent by the god Poseidon to terrorize the land of Ethiopia.

Sacrificing Andromeda to Cetus was the price Ethiopia's king Cepheus needed to pay as atonement for his wife Cassiopeia's boast that her daughter, Andromeda, was more beautiful than the nymphs of the sea. Poseidon could never forgive a boast like that. So, when Perseus appeared and approached Andromeda's grief-stricken parents, promising them that he would kill the monster in exchange for Andromeda's hand in marriage, they readily agreed.

As Cetus headed through the waves for Andromeda, Perseus, wearing winged sandals and armed with an unbreakable sword, attacked, swooping from the sky above. The hero and the monster battled fiercely, but Perseus escaped the monster's fangs and lashing tail, ultimately striking a fatal blow. The monster bled profusely—so much so that its streaming blood colored the Red Sea red. Perseus then turned the monster's massive corpse to stone by exposing it to the severed head of the Gorgon Medusa, a secret, lethal weapon that he had obtained in an earlier daring adventure.

When, after a long life together, Perseus and Andromeda died, the goddess Athena raised them to the heavens as constellations. And, so that the story of their heroic lives might always be remembered, some say she made Andromeda's parents and the monster Cetus constellations, too, though this commemorative act is also attributed to Poseidon, who wanted all participants in the Andromeda tale to be recognized. As a constellation, Cetus straddles the Celestial Equator. While Cetus's coiling body is located in Southern Quadrant 1, its head and neck extend into Northern Quadrant 1.

Best visibility: October to December (65°N–79°S)

See also the constellations Andromeda, Cassiopeia, Cepheus, and Perseus.

ERIDANUS

The River

He told how Phaethon, struck by a blast of heavenly fire, fell tumbling from the chariot headlong through the air, half-burned and buried in that Celtic river, and why, along the banks of Eridanus, his sisters moan the loss of that bold youth, tears streaming from their leaves.

———

Nonnus, *Dionysiaca* 38.91–95

MAIN CHARACTERS:
- Clymene, the mother of Helios
- Heliades, the daughters of Helios and sisters of Phaethon
- Helios, the Sun god
- Phaethon, the young son of Helios
- Zeus, king of the Greek gods

In Greco-Roman mythology, the constellation Eridanus represents the Eridanus River and is closely linked to the story of Phaethon, one of several memorable

SOUTHERN CELESTIAL QUADRANT I
Eridanus

charioteers associated with the constellation Auriga. Phaethon was the young son of Helios, the Sun god, and a nymph called Clymene. One day, a playmate ridiculed his claim to be the son of a divinity, and a now doubting and shaken Phaethon asked his mother for proof of his parentage. She advised that he learn the truth from Helios himself, and so Phaethon undertook the long journey from his home in Egypt to the Sun god's palace in the farthest East, this being the place from which, every morning, the Sun rises anew. The palace was unmistakable when he came upon it. It shone brightly with gold, bronze, and silver—almost too bright to behold. Inside the palace, Phaethon found Helios himself, who asked the reason for his visit. Phaethon's response was that he had come to ask a favor, and a favor—any favor he wished—was duly granted him. Helios's readiness to agree was, for a god, an uncharacteristic mistake. What Phaethon asked was to drive his father's blazing chariot, a thing that quickly proved too difficult for him to do. Phaethon lacked the skill and strength to rein in the chariot's horses, who veered upward high into the heavens, threatening to burn the constellations, and then plunged down again, scorching the earth, causing lakes to dry, and roasting people and their flocks. Zeus saw the need to intervene without delay, and hurled a thunderbolt at Phaethon, who, burned by Zeus's fire, tumbled from the airborne chariot into a river below. That river was known as the river Eridanus. Learning of their brother's fate, Phaethon's sisters, the Heliades ("daughters of Helios") gathered on the river's banks and wept, overcome by grief. Their sadness was so extreme that they turned into poplar trees that eternally shed tears of amber.

Where, exactly, this river is located was a mystery in antiquity and remains a mystery to this day. Ancient authors identified it with various European rivers, especially those in which amber could be found, among them what are now known as the Rhine, Rhone, Danube, and Po. However, the Nile, too, was identified as Eridanus, as was the fabled river Ocean that was believed to flow around the edge of a platter-shaped Earth. Indeed, many ancient authors referred to it simply as "the river." Whatever its actual physical location may be, a starry likeness of Eridanus flows forever through the heavens.

Best visibility: November to January (32°N–89°S)

See also the constellation Auriga.

SOUTHERN CELESTIAL QUADRANT I

LEPUS

The Hare

The hare is said to be running from the hunter Orion's dog for this reason: when people wanted to represent Orion as a huntsman, they showed a fleeing hare at his feet.

———

Hyginus, *Astronomica* 2.33.1

MAIN CHARACTERS:
Artemis, goddess of the hunt and protectress of wild animals
Gaia, the Earth goddess
Orion, one of classical mythology's greatest hunters
Zeus, king of the gods

Lepus (*the Hare*) darts across the night sky beneath the feet of the hunter Orion and is pursued by Orion's favorite dog. The dog is represented by the constellation Canis Major (*the Greater Dog*), which lies in close proximity to its master.

The hare is said to have been placed by Zeus in the heavens with Orion as a means of illustrating that this hero was a hunter. In fact, Orion was one of classical mythology's most skilled hunters, but he made the fatal mistake of boasting of his skill in a way that deeply offended the gods, especially Artemis: Orion claimed that he could hunt and kill every animal on the island of Crete. Artemis, whose company Orion kept while on the island, was the goddess of the hunt but also the protectress of animals. Her realms of influence might today seem contradictory, but in antiquity they were not. Hunters then were keenly aware that they were taking another life and considered an animal life taken to be a sacred gift. For this gift, they asked the animal's forgiveness, and shared their bounty with the gods through the ritual of sacrifice. Killing sprees were not condoned.

For this boast, Orion was duly punished. The gods, either Artemis or the Earth goddess Gaia, sent an enormous scorpion to kill him. The scorpion became the constellation Scorpius, by will of Zeus who wished to warn humanity to value their fellow beasts, but Orion became a constellation, too, and along with him the hare and one (or both) of his dogs. If both, then they are the Greater and Lesser Dogs, Canis Major and Canis Minor.

There is an alternate story for the presence of the hare in the sky, however, and it does not feature Orion. According to this lesser-known variant, a young man who had a particular fascination with rabbits and their great fertility introduced one pregnant rabbit to the Greek island of Leros. When the rabbit gave birth, the other citizens of Leros took an interest in this new animal, too, coveting them as gifts or novelties for sale. Soon everyone was raising rabbits, and the island became overrun with them. The rabbits ate the people's seeds and crops, so that starvation became rampant. With great difficulty, the rabbits were eventually driven from the island, and the people saw to it that an image of a rabbit be placed among the stars as a warning to other people: "Beware of that which you covet, for it may bring more harm than good."

Best visibility: January (62°N–90°S)

See also the constellations Canis Major, Canis Minor, Orion, and Scorpius.

SOUTHERN CELESTIAL QUADRANT I
Lepus

SOUT
CELE
QUAD

HERN
STIAL
RANT
II

SOUTHERN CELESTIAL QUADRANT II
Argo Navis

SOUTHERN CELESTIAL QUADRANT II

ARGO NAVIS

The Ship Argo

When Argo's beak plowed the wind-swept seas and her oars churned up frothing waves, the Nereids, nymphs of the sea, rose up from the deep amid the snow-white foam, wondering at this new, strange thing. On that day, and on no other since, mortals could see the sea-Nymphs, their heads, shoulders, and bare breasts raised above the waves.

Catullus, *Carmina* 64.12–18

MAIN CHARACTERS:
- Achilles, the Greek hero known for his exploits during the Trojan War
- Aeetes, king of Colchis and father of Medea
- Aeson, former king of Iolcus and father of Jason
- Argonauts, crew of the ship *Argo*
- Argus, creator of the ship *Argo*
- Athena, goddess of defensive war and patron goddess of craftspeople
- Castor, one of the twins known as the Gemini
- Charybdis, a massive whirlpool
- Chiron, a noble and learned centaur
- Circe, a sorceress and aunt of Medea
- Harpies, monstrous bird-women who snatched away Phineus's food
- Helios, the Sun god
- Hera, queen of the gods
- Hercules, the most beloved of all Greek heroes and known for his Twelve Labors
- Jason, prince of Iolcus and captain of the *Argo*
- Medea, princess of Colchis and a sorceress
- Orpheus, ancient Greece's most skilled musician
- Pelias, illegitimate king of Iolcus
- Phineus, a man tortured by the Harpies and rescued by the Argonauts
- Pollux, twin brother of Castor
- Scylla, a monster with a woman's head, a waist growing dogs' heads, and a fish tail

CLASSICAL MYTHOLOGY OF THE CONSTELLATIONS

Talus, a near-invincible giant of bronze
Zeus, king of the gods

The large constellation Argo Navis (*the Ship* Argo) is named after what was reputedly the first seafaring ship ever to be constructed. Understandably, the sight of this vessel, a wholly unfamiliar thing, was completely astonishing to the gods and creatures of the sea when they first laid eyes upon it. *Argo* was built by the skilled craftsman Argus with guidance from the goddess Athena, who supplied the ship with a magical beam crafted from the sacred oak tree of Zeus at Dodona. This oak could utter prophecies, conveyed through the rustling of its leaves, and the beam, too, could speak, offering guidance to the new ship's crew. The *Argo*'s captain was Jason, rightful heir to the throne of Iolcus in Thessaly, northern Greece, and the distinguished crew consisted of the best and bravest men from all of Greece. Among them were the father of Achilles, the renowned singer-poet Orpheus with his famous lyre (Lyra), Hercules, and the munificent twins Castor and Pollux (Gemini). Together, the captain and crew were known as the Argonauts, "sailors on the *Argo*," and their journey was an epic one.

The tale begins with Jason, son of Aeson, Iolcus's rightful king. Aeson was driven from the throne by his half-brother Pelias, who would have killed the newborn baby Jason had his quick-thinking mother not pretended that he was stillborn. Secretly, she sent the baby Jason from the kingdom to the mountain cave where the centaur Chiron (Centaurus) lived. There, in Chiron's care, Jason was raised and educated. Quite reasonably, when he reached maturity, Jason wished to return home, hoping to claim the throne that was rightfully his.

On his way, Jason came to a raging river, and on its banks stood an old woman, too feeble to make the crossing on her own. Moved by her predicament, Jason offered to carry her across on his shoulders, a favor that she gratefully accepted. Both crossed safely, the only mishap being the loss of one of Jason's sandals to the river's currents. The old woman, as it happened, was none other than the goddess Hera in disguise, and this had been a test for the young man. It was a test that he passed with flying colors. Consequently, Hera would help him in the course of his many adventures from this day on.

At long last, Jason arrived at Iolcus and requested an audience with King Pelias. Pelias, meanwhile, had learned from a prophecy that he should beware of the arrival of a stranger wearing just one sandal. The appearance of Jason, therefore, unnerved him. When Jason revealed his identity, Pelias readily agreed to give up the throne, but only on one condition: Jason first had to bring him the golden fleece of a sacred ram (Aries) that was guarded by a sleepless dragon in the

SOUTHERN CELESTIAL QUADRANT II

kingdom of Colchis on the distant shores of the Black Sea. This, Pelias expected, was a task that Jason would never be able to survive, much less complete. Jason, of course, agreed and prepared to make this perilous journey. A ship was built—the *Argo*—and an illustrious crew assembled. The journey was far from easy, and the Argonauts faced countless dangers along the way.

First they came upon a group of women living on the island of Lemnos who had killed every man as well as every male child on the island, and now wanted the Argonauts for husbands. They also encountered the Bebryces, a hostile tribe whose king would challenge all newcomers to a boxing match, a match that he would inevitably win. His match with Pollux, who killed him with a single blow, proved to be his last.

Along the way, the Argonauts chanced upon Phineus, a son of the god Poseidon but, in spite of his divine parentage, a man enduring endless torture. Phineus had been blinded by Zeus, who was angered at his skill in prophecy, and all food that he tried to bring to his lips was snatched from his hands by the dreaded Harpies, birdlike female monsters. What food the Harpies did not manage to grab they fouled with their droppings. Out of pity for his plight, the Argonauts drove the Harpies off, and a grateful Phineus revealed the only means by which they might safely pass between the deadly Clashing Rocks that controlled passage from the Aegean into the Black Sea. If a dove could safely pass between these rocks, so could the *Argo*. One of the Argonauts had a dove, and it was able just to make it through.

When the Argonauts arrived on the shores of Colchis, that region's king, Aeetes, received them hospitably but feared that their purpose was not just to take away the golden fleece but also to strip him of his kingdom. The best course of action, Aeetes thought, was for him to set Jason a further series of impossible tasks. Aeetes's expectation was that Jason would be killed, saving him from having to resort to murder and thereby garnering the disfavor of the gods. Jason was to yoke a pair of fire-breathing bulls and then to plow a field, sowing dragons' teeth that would sprout not plants but fierce warriors. This accomplished, the fleece would be his.

Fortunately, Jason had the help of Aeetes's young daughter Medea, a sorceress who had fallen in love with him at first sight. She provided a salve that would protect Jason's skin against the bulls' flaming breath, and she told him how to survive the warriors springing from the dragons' teeth: he was to throw a stone into their midst, and, in confusion, they would turn upon each other. These tasks accomplished, Jason returned to the palace to claim his prize, but Aeetes would not release it. Again, Medea came to Jason's aid. With a potion, she put the dragon that guarded the fleece to sleep, allowing the Argonauts to grab the fleece and make a swift escape. The Colchians sailed in pursuit, but the *Argo* evaded them.

The journey home was no less perilous than that to Colchis. The Argonauts came upon the sorceress Circe, the bronze giant Talus, and the monsters Scylla and Charybdis, among others. When Jason finally returned to his home at Iolcus, Pelias, too, refused to honor his promise. He would not give up the throne. Once more Medea, now Jason's wife, came to his aid. She convinced Pelias's daughters to allow her to rejuvenate the old king and demonstrated her abilities with an elderly ram that she chopped up and placed in a brew. The ram leaped out alive, rejuventated, and good as new. With Pelias's daughters' blessing, Medea chopped up the king and placed him in a cauldron filled not with a magic brew but only with boiling water. This spelled Pelias's end, but Medea's purpose was not achieved. Jason could not now become king because his wife had committed murder. In the Greek world, the punishment for murder was death or exile. So, Jason and Medea fled to the kingdom of Corinth.

In Corinth, Jason made a fatal mistake. He decided to wed the princess of that realm. In a jealous rage, Medea contrived to cause him as much pain as possible. She gave her children by Jason a crown and cloak to give to Jason's new bride. These gifts had been poisoned, and caused the princess and her father, who ran to her rescue, to be consumed by fire. Medea then turned upon her own children, stabbing them to death one by one. A granddaughter of the Sun god Helios, Medea made her escape in that god's chariot with Jason left a broken man.

While Jason's end was far from honorable, his faithful ship the *Argo* was placed in the sky by the goddess Athena. In the sky, the *Argo* is shown from the rear and as if sailing backward. The constellation *Argo* known in antiquity was enormous, being almost 30 percent larger than Hydra. It was so large that later astronomers, thinking that it consisted of too many stars to keep track of easily, took measures to separate it into three constellations: Carina (*the Keel*), Puppis (*the Stern*), and Vela (*the Sails*). The separation was first made by French astronomer Nicolas-Louis de Lacaille in the eighteenth century and formalized by the IAU (International Astronomical Union) in 1930. Puppis and Vela lie to the north of Carina, all three in the Milky Way.

Best visibility: Carina, January to April (14°N–90°S); Puppis, January to February (39°N–90°S); Vela, February to April (32°N–90°S)

See also the constellations Aries, Centaurus, Gemini, Hercules, and Hydra.

SOUTHERN CELESTIAL QUADRANT II

CANIS MAJOR

The Greater Dog

> But when Erigone, the daughter of Icarius, longing for her absent father, realized he had not returned and prepared to go in search of him, Icarius's faithful dog, Maera, came to her, howling as if lamenting its master's death.
>
> Hyginus, *Astronomica* 2.4.4

MAIN CHARACTERS:
- Artemis, goddess of the hunt and protectress of wild animals
- Aura, goddess of the breezes
- Cephalus, a hunter loved by Eos and married to Procris
- Dionysus, the god of wine and the vintage
- Eos, goddess of the dawn
- Erigone, the devoted daughter of Icarius
- Europa, a Phoenician princess abducted by Zeus
- Hera, queen of the gods
- Icarius, a devout Athenian farmer
- Laelaps, a miraculous dog who always caught its prey
- Maera, the faithful dog of Icarius the farmer
- Medusa, a snake-haired Gorgon
- Minos, a son of Europa and king of Crete
- Orion, a legendary hunter
- Perseus, slayer of Medusa
- Procris, an Athenian princess loved by Minos and married to Cephalus
- Zeus, king of the gods

Classical mythology identifies several dogs with that pictured by the constellation Canis Major (*the Greater Dog*). Among them are Maera, the faithful dog of Icarius, a devout and generous farmer. The dog Maera, along with her human companions Icarius and Icarius's daughter Erigone, are known for their exemplary deeds in life and for their tragic deaths. Icarius and Erigone were among those select individuals who welcomed the god Dionysus when he came to Greece from the Near East. In the eyes of some, Dionysus was a threatening, polarizing figure.

CLASSICAL MYTHOLOGY OF THE CONSTELLATIONS

He was effeminate and foreign, a shape-shifter, and a god who broke down social barriers (between old and young, free and enslaved, male and female, even human and animal). Dionysus came to Greece bearing the gift of wine, which offered at least temporary release from hardships and worries to those who embraced the god. Icarius was rewarded for his faith and devotion by being instructed in the vintner's art, and, unselfish as Icarius was, he set out with his oxcart heaped with wineskins to share with all his neighbors. The first of these that he encountered were shepherds, and they were keen to sample Icarius's drink. They drank liberally and in excess, becoming first stupefied and then paranoid. Suddenly, they saw Icarius as a threat—surely he had tried to poison them! And so these men set upon the poor farmer in the dead of night and beat him to death with their clubs. Meanwhile, Erigone anxiously awaited her father's return, and grew more worried with passing time. Then, just as she resolved to leave the house in search of her dear father, his dog Maera appeared, howling terribly. Maera led Erigone to her father's body. Losing hope utterly and overcome with grief, Erigone hanged herself from a nearby tree, and a heartbroken Maera, faithful to the end, leaped into a well to her death. In recognition of their inherent goodness, the gods raised all three to the heavens as constellations: Icarius as the ox driver Boötes, Erigone as Virgo, and Maera as Canis Major.

Another dog associated with Canis Major is the great hunter Orion's favorite dog, an identification that makes good mythological sense particularly because Canis Major is situated near the constellation Orion. There is not much of a story associated with this dog per se apart from accompanying Orion on the hunt—in life as well as in the "afterlife" as a constellation. Orion's skill as a hunter was a source of the hero's fame but also the cause of his downfall. Overly filled with confidence, Orion boasted that he could successfully hunt and kill all the animals on the island of Crete. This boast filled the gods with alarm, especially Artemis, who once had cherished Orion's companionship. So, an enormous scorpion was sent to kill him by Artemis or the Earth goddess Gaia or by both, and when the scorpion's task was accomplished, Zeus raised it to the heavens as a constellation (Scorpius), as a reward for it and a warning to humans who risked overstepping their limits. Orion, accompanied by his dog, was raised to the heavens, too. This was the wish of Artemis, who wanted his heroic deeds to be remembered.

There is a further, especially strong candidate for Canis Major: the infallible hunting dog Laelaps. This dog's hunting instincts were so finely honed that no prey could escape him. Laelaps had been given by Zeus to Europa, a Phoenician princess, to serve as her guard dog. The story of Zeus's infatuation with Europa and his abduction of her is well known. Zeus transformed himself into a lovely,

SOUTHERN CELESTIAL QUADRANT II
Canis Major

tame white bull in order to approach her. An unsuspecting Europa, charmed by this animal, climbed onto its back, and no sooner had she done this than it (Zeus in disguise) bolted for the sea and swam, carrying her, all the way to Crete. Zeus and his abduction of Europa are commemorated by the constellation Taurus (*the Bull*).

By Zeus, Europa became the mother of Minos, Crete's most powerful king, and it was he who inherited his mother's dog. Minos later gave this valuable dog to Procris, a princess of Athens, when she cured him of an ailment that had mysteriously beset him. By some accounts, Minos was in love with Procris, too, and hoped to win her affection with the gift of Laelaps and a javelin. Procris, however, loved a young man named Cephalus and became his bride. Although they were faithful and devoted to each other, theirs was not a happy tale. Procris gave Laelaps and her spear to Cephalus, an ardent hunter, to take with him into the wild. While out one day hunting, Cephalus caught the eye of Eos, goddess of the dawn, who carried him off and bore a son by him. In the course of time, Eos did let Cephalus return to Procris, but Procris was understandably concerned every time he left the house. Would he be meeting some other lover? Her suspicions appeared to be confirmed by a rumor: Cephalus had been heard whispering to a woman called Aura. So, Procris followed him into the woods one day and saw him resting from the heat and calling to the cooling breeze, *aura*, to come refresh him. Aura means "breeze" in Greek and Latin. Procris realized her mistake and tried to withdraw as quietly as possible from her hiding place. It was not quietly enough. The bushes rustled as she moved, and Cephalus, thinking she was a deer, killed her with his javelin.

The unfortunate Cephalus later lent Laelaps to Amphitryon, a son of the great hero Perseus, famed for beheading the snake-haired Gorgon Medusa. Amphitryon, as it happened, was setting out to hunt a particular fox that had been sent by the goddess Hera to wreak havoc in the countryside of Thebes. Hera had ensured that this fox could never be caught, and since Laelaps always caught his prey, their chase was destined to be endless. Zeus intervened, turning both animals to stone. This legendary race was then memorialized in the stars, with Laelaps as Canis Major and the fox as Canis Minor (*the Lesser Dog*).

The prominent constellation Canis Major contains the brightest of all stars and one of the most culturally significant in classical antiquity: Sirius, known as "the Dog Star," also called Canicula ("little dog") in Latin. The appearance of this very bright star in the late-summer night sky signaled the onset of great heat in the Greco-Roman/Mediterranean world and thus danger to people and their flocks.

Canis Major is located near the constellations Orion and Canis Minor, and both dogs (Canis Major and Canis Minor) appear to follow Orion over the night sky.

Best visibility: January to February (56°N–90°S)

See also the constellations Boötes, Canis Minor, Orion, Perseus, Taurus, and Virgo.

CRATER

The Cup

Astronomers of old drew the crater in the stars, so that humankind would remember that no person can do an evil deed with impunity, nor are acts of violence forgotten.

Hyginus, *Astronomica* 2.40.3

MAIN CHARACTERS:

    Apollo, the god of prophecy, medicine, music, and archery
    Demophon, a king of Elaeusa
    Hercules, the greatest of all Greek heroes and known for his Twelve Labors
    Matusius, a nobleman of Elaeusa
    Pholus, a kind and devout centaur
    Zeus, king of the Greek gods

The relatively small and faint constellation Crater, commonly known as "the Cup," illustrates how the names of constellations could shift and become confused even in antiquity. Some of the earliest Greek sources call this constellation *kantharos*, *kalpe*, and *kylix*, all of which mean "cup." Some Roman authors, too, called the constellation "cup." Crater (Greek *krater*), on the other hand, actually means "a mixing bowl," specifically a mixing bowl for wine and water, not a vessel to be drunk from directly. The Greeks never drank their wine undiluted; doing so was a sign of dissolution and lack of self-control. Wine was mixed with water in a *krater* and then poured by pitcher into cups.

It happens to be the case that cups and mixing bowls had a similar shape in antiquity, though mixing bowls were larger. The vessel in the sky could reasonably

SOUTHERN CELESTIAL QUADRANT II
Crater

be either, and the stories told in antiquity could be about either as well. Whether cup or mixing bowl makes little difference to the narratives. Somehow, for some unknown reason, the name Crater ultimately prevailed in the course of time, but the constellation is still thought of today as a cup.

The Cup is typically linked with the story of Corvus (*the Crow*) and goes as follows. In preparation for a feast in honor of Zeus, the god Apollo wanted to make a ritual offering of pure water. Apollo sent a crow, his sacred bird, to fetch water for him from a nearby spring. It was a simple errand that took longer than it should have because the crow became distracted by a fig tree densely hung with fruit. The figs were not quite ripe yet, however, so the crow decided to wait. Days later, when it had eaten its fill of ripened fruit, the crow returned to Apollo with an empty cup in its beak and a water snake in its talons. This snake, he lyingly reported, had kept him from the spring. Knowing that the truth was different, an angry Apollo turned this white crow's feathers black and then placed a starry picture of the crow, the snake, and the cup in the heavens as a warning to those who defied the gods. The crow is near the cup but remains eternally thirsty because this heavenly cup is closely guarded by the snake.

A second tale that is sometimes associated with Crater is that of the kind and god-fearing centaur Pholus, who was accidentally and fatally wounded by one of Hercules's poisoned arrows. According to this story, Pholus was commemorated by the gods with the constellation Centaurus (*the Centaur*), and his mixing bowl, too, was lifted to the heavens. An exemplary host while he was still alive, Pholus offered Hercules some of his special wine. This wine attracted the neighboring centaurs, who tried to steal it. Centaurs were notorious for loving wine, drinking in excess, and becoming violent. Hercules drove them off, but Pholus met his untimely end shortly afterward when handling one of Hercules's arrows and dropping it on his foot.

A third version of the cup's appearance in the night sky centers on a king of Elaeusa, a city near Troy. This city was suddenly beset by the plague, and many people died. The distraught king, Demophon, sent an envoy to consult the oracle of the god Apollo to ask what remedy there might be for the ongoing deaths. The oracle's response was that the people of Elaeusa would have to sacrifice the daughter of a noble family annually in order for the plague to dissipate. The king had several daughters and, of course, did not want any of them to die. So, he told his subjects that the victim would be chosen by lot. What he did not tell them was that his daughters' names were never entered in the drawing.

One nobleman did discover the king's deceit and exposed it. This caused the king to have the nobleman's daughter killed without the usual lottery. For a time, her father pretended that this was simply fate and that he bore the king no ill will. In secret, however, he plotted his revenge. This man, Matusius by name, invited the king and his family to a celebration. The king's daughters, he said, should arrive first. At Matusius's home, the king's daughters were killed in an ambush and their blood mixed with water in the ceremonial "crater". After enjoying a cup of what he thought was wine, the king asked where his daughters were. In shock and grief at the response, he seized Matusius and threw him and the crater into the sea. This crater, then, appeared among the stars as a cautionary tale for humans intent on evil doing.

Best visibility: April (65°N–90°S)

See also the constellations Centaurus, Corvus, and Hercules.

HYDRA

The Water Snake

Hercules killed the dreaded Hydra, disemboweled her, and dipped his arrows in her gall. So it came to pass that whatever he later hit with his arrows could not escape death.

Hyginus, *Fabulae* 30.3.1-5

MAIN CHARACTERS:
Bellerophon, the hero who tamed Pegasus and slew the Chimera
Cancer, an enormous crab that attacked Hercules
Cerberus, the triple-headed dog that guarded the Underworld
Chimera, a goat-lion-serpent-hybrid monster and sibling of the Hydra
Echidna, the Hydra's half-serpent and half-human mother
Hera, queen of the gods
Hercules, greatest of the Greek heroes and famous for his Twelve Labors
Hydra of Lerna, a many-headed serpent and adversary of Hercules
Leo, the Nemean Lion killed in Hercules's first labor and sibling of the Hydra
Pegasus, a winged horse and offspring of Medusa
Typhon, the Hydra's monstrous father who had one hundred flaming
    snake heads

SOUTHERN CELESTIAL QUADRANT II
Hydra

CLASSICAL MYTHOLOGY OF THE CONSTELLATIONS

The infamous Hydra (*the Water Snake*) was an enormous female water snake that lived in the marshes of Lerna in the eastern Peloponnese. The Hydra, one of the fearsome offspring of the snake-monsters Typhon and Echidna, had many heads, some ancient authors stating that there were as many as one hundred, and her venom, gall, blood, and breath were poisonous. As one might expect, her siblings included other monsters. These were the ferocious lion (Leo) that prowled the countryside of Nemea, the triple-headed dog Cerberus that guarded the Underworld, the dragon (Draco) who guarded the Hesperides' golden apples, and the Chimera, the goat-lion-serpent hybrid monster slain by the hero Bellerophon, famed for taming the winged horse Pegasus.

As with the Nemean Lion, the goddess Hera nurtured the Hydra to terrorize the inhabitants of the surrounding countryside and ultimately to become one of the hero Hercules's adversaries. As the second of his Twelve Labors, Hercules was sent to kill the Hydra. This was an especially challenging task because, as Hercules discovered, two new heads grew in place of each that he lopped off. And, as a further challenge, an enormous crab (Cancer) emerged from the swamp to help the Hydra in her struggle with Hercules. This crab Hercules simply stomped to death, but the Hydra, of course, remained alive.

As it happened, Hercules's nephew Iolaus, his companion on this adventure, was carrying a torch, which the hero grabbed and used to quickly cauterize each freshly severed neck. One of the Hydra's heads, however, proved to be immortal, but Hercules found a solution to this challenge, too. He buried the severed, deathless head, covering the site with a massive boulder. Then Hercules set to cutting up the Hydra's corpse in order to harvest those poisonous fluids that would make his arrows fatal, even if only lightly grazing their target.

Among the many victims of Hercules's poisoned arrows were the centaurs Chiron, Nessus, and Pholus—all three being centaurs associated with the constellation Centaurus—and even Hercules himself. Hercules's arrows, too, were memorialized in a constellation, Sagitta. As for the Hydra and her friend the crab, they were placed among the stars by Hera. Hydra, the largest of all currently recognized constellations, winds her way through the heavens, over three celestial quadrants. She extends her body through Southern Quadrants 2 and 3 and raises her head into Northern Quadrant 2.

Best visibility: February to June (54°N–83°S)

See also the constellations Cancer, Draco, Hercules, Leo, Pegasus, and Sagitta.

SOUT
CELE
QUAD

HERN STIAL RANT III

SOUTHERN CELESTIAL QUADRANT III
Ara

SOUTHERN CELESTIAL QUADRANT III

ARA

The Altar

It is believed that the gods first made offerings on this altar when they formed an alliance to oppose the Titans. The altar was crafted by the Cyclopes. From this observance humans established the custom to make sacrifice before undertaking any new venture.

Hyginus, *Astronomica* 2.39.1

MAIN CHARACTERS:
Chiron, a noble and gentle centaur
Cronus, the youngest of the Titan gods
Cyclopes, one-eyed giants
Gaia, the Earth goddess
Hecatoncheires, monsters with one hundred hands
Hercules, the greatest of all Greek heroes and known for his Twelve Labors
Uranus, god of the heavens and spouse of Gaia
Titans, twelve siblings born to Gaia and Uranus
Zeus, a son of Cronus and later king of the gods

The constellation Ara (*the Altar*) was believed to represent the altar at which the god Zeus and his siblings made offerings of incense and poured libations to seal an alliance between them. The joint venture that they planned to undertake was a war against the older Titan gods. What follows is the altar's story.

This new war of Zeus's was just another, though significant, conflict between various generations of the gods. The king of the Titans, Cronus, was also the youngest of them. He had become their leader by being the only one to dare oppose Uranus, their father. The Earth goddess Gaia, Uranus's spouse, had given birth not only to the twelve Titans but also to six monsters: three Cyclopes, each with a single eye, and three Hecatoncheires, each with one hundred hands. Uranus detested the monsters, and pressed them back into their mother at the moment of their birth. A distressed Gaia asked her other children for help, and only Cronus came forward. He then attacked his father and replaced him as king of the gods.

In the course of time, Cronus married his Titan sister Rhea, but he learned from a prophecy that one of his children would overthrow him, just as he himself had overthrown his own father. To prevent this from happening, Cronus swallowed each of his six children as they were born. By the time the sixth emerged from her womb, Rhea had formulated a plan to outwit her husband: she handed him not her youngest-born, Zeus, but rather a stone in swaddling clothes. Cronus did not notice the difference.

The infant Zeus, meanwhile, was sent to the island of Crete to be raised, and when fully grown, he gave his father an emetic that caused him to disgorge his siblings. With Zeus taking the lead, Cronus's children banded together to make war on their father, and before opening hostilities, they performed a ritual offering at a particular, special altar that had been crafted by the Cyclopes. The fact that an eagle appeared in the sky above the smoking altar as Zeus was making sacrifice was interpreted as a good omen: their punitive assault on the Titans would succeed. The battle of the Titans and the Olympian gods, as Zeus and his siblings were called, left the Olympians victorious. Zeus now was king of the gods, and the offering he made became precedent for the Greeks and Romans alike before undertaking any venture. As for the altar and the eagle (Aquila) that featured so prominently in this story, these were commemorated by the gods with constellations in their name.

As with so many other constellations, there is an alternate story associated with Ara. That tale also involved Zeus, but the altar in question was a different one. The constellation Ara is located along the Milky Way near the constellations Centaurus (*the Centaur*) and Lupus (*the Wolf*), and this variant tale takes the relative position of these three constellations into consideration. When the noble centaur Pholus was accidentally killed by one of Hercules's arrows, Zeus took pity on him, especially since he had been extremely pious and god-fearing. In order to memorialize Pholus's piety, he placed an altar (Ara) nearby him in the heavens as well as an animal, a wolf (Lupus) or some other wild beast, for him to sacrifice.

Best visibility: June to July (22°N–90°S)

See also the constellations Aquila, Centaurus, and Lupus.

SOUTHERN CELESTIAL QUADRANT III

CENTAURUS

The Centaur

Assuming the form of a stallion, Cronus lay with Philyra, daughter of Oceanus. By him she bore Chiron the Centaur, who is said to have been the first to invent the art of healing.

Hyginus, *Fabulae* 138.1.1

MAIN CHARACTERS:

Achilles, a Greek hero famed for his exploits during the Trojan War
Castor and Pollux, twin brothers of Helen of Troy, known as Gemini
Chiron, the gentlest and most highly cultured of all centaurs
Cronus, one of the ancient Titan gods and father of Chiron
Deianeira, wife of Hercules
Dionysus, Greek god of wine
Hercules, the greatest of all Greek heroes
Ixion, a legendary Greek sinner and ancestor of Nessus
Jason, the Greek hero who traveled on the ship *Argo* to fetch the famed golden fleece
Nessus, an unruly and beastly centaur who attempted to rape Hercules's wife
Nephele, a cloud in female form and mate of Ixion
Philyra, daughter of Oceanus and mother of the centaur Chiron
Pholus, a gentle centaur killed accidentally by one of Hercules's arrows
Zeus, king of the Greek gods

The constellation Centaurus (*the Centaur*) is named after a particular centaur, a hybrid creature that was half-human and half-horse, the human half being the head and torso. Exactly what centaur this constellation commemorates was a matter of debate even in antiquity. There are three candidates, all of them featuring prominently in classical mythology: Chiron, Nessus, and Pholus. Of these, Chiron was by far the most famous, but all three will be considered here.

Generally speaking, centaurs were unruly creatures with a reputation for drunkenness and violence, especially toward women. Chiron was different. Instead, he was not only gentle but also highly cultured, so much so that, as children, Greece's foremost heroes were sent to live with him in his cave to be

CLASSICAL MYTHOLOGY OF THE CONSTELLATIONS

educated. Among them were the great Achilles, best known for his exploits in the Trojan War; Hercules, who famously endured twelve horrific labors; the famous twins (Gemini) Castor and Pollux; and Jason, who sailed to the far reaches of the known world to retrieve the fleece of a very special golden ram (Aries). Skills and arts in which Chiron excelled included archery, music, sculpture, and medicine. By some accounts he was the first to collect herbs to be used in healing, thus becoming the inventor of medicine. As it happens, all of his pupils aside from Achilles became or had direct links to constellations.

By some accounts, Chiron was the son of Cronus, one of the very ancient Titan gods, and the lovely sea-nymph Philyra, a daughter of Oceanus, god of the great river believed to flow around the edges of a platter-shaped Earth. Cronus transformed himself into a horse when he lay with Philyra, hoping in this way to escape detection by his wife and sister, Rhea. This, then, is the explanation for Chiron's hybrid form. As for his gentle nature, this, too, had to do with his parentage, for the other, wilder and unruly centaurs were descended from the cloud-woman Nephele and Ixion, a legendary sinner who attempted to seduce Zeus's wife, the goddess Hera. Learning of Ixion's interest in Hera, Zeus placed Nephele, a shape-shifting cloud-deity impersonating Hera, in his bed to find out if this was true. It was. As punishment for his affront to Zeus, Ixion was fastened to a flaming wheel that spun endlessly in the Underworld after his death.

In addition to his claim to raising the best-known heroes of Greece, Chiron was notable for the manner of his death, or, more properly, the end of his Earthly existence. As the child of immortals, he could not actually die, but a time came when he wished desperately that this were not the case. On one of his many adventures, Chiron's former pupil Hercules visited him and accidentally wounded him with one of his poisoned arrows. Chiron's pain was so great that the gods took pity on him and let him quit the Earth and Earthly suffering, raising him instead to the stars.

Nessus, the second reputed namesake of the constellation Centaurus, was a wholly different creature from Chiron in temperament and manners. He is known primarily as the cause of the great hero Hercules's gruesome death. Nessus offered to carry Hercules's wife Deianeira across a dangerous, swollen river, and a weary Hercules was grateful, initially, for the offer. However, once Deianeira was on his back, the centaur became uncontrollably aroused and attempted to gallop off with her. Just then, Hercules was still midstream but heard Deianeira's cries. Turning quick as he could, he shot the centaur with his bow and arrow, but Nessus did not die immediately. He had just enough time to offer Deianeira some of his blood to be used as a love potion if ever she felt that Hercules no longer loved her. The

SOUTHERN CELESTIAL QUADRANT III
Centaurus

potion was no charm but, instead, a poison that an unsuspecting Deianeira later applied to Hercules's cloak. His skin began to burn and peel, the pain excruciating. Hercules would end his own life by climbing onto a burning funeral pyre.

The third centaur considered by some ancient authors to be the source of the constellation Centaurus is Pholus, a relatively obscure character. In fact, both he and even Chiron were believed by some to have become the constellation Sagittarius, a centaur wielding a bow and arrow. In any event, Pholus was, like Chiron, an unusually gentle centaur who lived in a mountain cave. It was said that Hercules also visited Pholus in the course of one of his many adventures. As any good host would do, Pholus offered his guest some of the precious wine that the wine-god Dionysus had entrusted him with keeping. The very scent of this potent drink attracted neighboring centaurs and drove them mad. These intoxicated centaurs set upon Hercules, who drove them off using his bow and arrow. One arrow fell on Pholus's foot as he was handling it, a fatal accident, as the arrow had been dipped in the Hydra's poisonous blood. Hercules then buried him and, in honor of his friend, named the mountain on which he had lived Pholoe. The gods, in turn, made him a constellation.

The large constellation Centaurus depicts a centaur—whether Chiron, Nessus, or Pholus—holding a spear in his hands.

Best visibility: April to June (25°N–90°S)

See also the constellations Argo Navis, Crater, Gemini, Hercules, Hydra, and Sagittarius.

CORVUS

The Crow

Distraught, Apollo seized his unborn child from the flames, pulling him from his mother's corpse, and brought him to the cave of Chiron, that famous hybrid creature. Then he summoned the crow, fool that it was expecting a reward for the truth it had revealed. Instead, the god excluded it from flocks of birds whose feathers were a pure white.

Ovid, *Metamorphoses* 2.628–632

SOUTHERN CELESTIAL QUADRANT III

MAIN CHARACTERS:
    Apollo, the god of prophecy and medicine as well as father of Asclepius
    Asclepius, a son of Apollo who became a healing deity
    Chiron, a noble and highly cultured centaur
    Coronis, the mother of Asclepius by Apollo
    Hydra, a many-headed water serpent killed by Hercules
    Ischys, the mortal lover of Coronis
    Zeus, king of the gods

The constellation Corvus (*the Crow*) appears in two different myths featuring Apollo, the god of medicine, prophecy, music, and archery. The first of these centers on Coronis, a princess of Larissa in northeastern Greece, who was exceptionally beautiful. Apollo became her lover, and, fearful that she might stray into the arms of someone else, he ordered a crow to guard her when he was away. Coronis did stray, beginning an affair with a handsome young man named Ischys. The crow reported Coronis's faithlessness to Apollo, who in a jealous rage shot Coronis with an arrow. As she lay dying in his arms, Coronis revealed that she was pregnant with Apollo's son. It was too late for Apollo to save Coronis, although he was filled with regret for killing her. He did just have time to save his unborn son, however, removing him quickly from his mother's body as her funeral pyre was lit.

    Although the crow had only done what it was told, and was innocent of any actual misdeed, a grief-stricken, angry Apollo changed its white feathers to black. The crow would now be a bird of ill omen. Interestingly, it is not the actual crow but rather Coronis who was commemorated with the constellation Corvus. Coronis's name, as it happens, means "crow" in Greek, and while a crow was complicit in her downfall, she had only herself to blame. As for her son, he grew up to become the famous physician and healing god Asclepius, who was himself honored with the constellation Ophiuchus (*the Serpent-Holder*). The noble centaur Chiron, who had instructed Asclepius in the art of healing, was raised to the night sky, too. The latter was memorialized by the constellation Centaurus (*the Centaur*).

    The second, lesser-known myth explaining the origins of the constellation Corvus is based on its proximity to the constellations Hydra (*the Water Snake*) and Crater (*the Cup*). According to this story, Apollo wanted to pour a libation (drink offering) before a feast in honor of Zeus and sent a crow, his sacred bird, to a spring to fetch some water for him. Near the spring, the crow saw a grove of fig trees whose fruit had not yet ripened. Perching on one of these trees, it decided to wait until the figs were ripe and it could have a tasty meal. A number of days passed before this happened. Then, having eaten its fill, the crow returned, not

SOUTHERN CELESTIAL QUADRANT III
Corvus

with a cup full of water but with a water snake in its talons. Apollo, meanwhile, had given up on the crow and used some other, less pure water for his ritual. Annoyed at the crow—not only because it was tardy but also because it had lied, saying the snake had kept it from reaching the spring—he made it so that the crow would always go thirsty during the period when figs were ripening. Since Apollo was the god of prophecy, he knew well that the bird had not told the truth. And as a warning to those who disobeyed the gods, Apollo placed the crow in the heavens with a cup of water (Crater) beneath it, hard to reach and guarded by a water snake (Hydra) whose tail the crow is shown pecking at.

Best visibility: April to May (65°N–90°S)

See also the constellations Crater, Hydra, and Ophiuchus.

LIBRA

The Scales

This sign of Scorpius is divided into two on account of the enormous size of its main parts. One part of it—the claws—some of our writers have called by a different name: the Scales.

———

Hyginus, *Astronomica* 2.26.1

MAIN CHARACTERS:
 Artemis, goddess of the hunt and protectress of wild animals
 Astraea, a goddess of justice
 Dikē, a goddess of justice
 Gaia, the Earth goddess and "mother" of Orion
 Hermes, the messenger god and a father of Orion
 Orion, one of classical mythology's greatest hunters
 Poseidon, god of the sea and one of the fathers of Orion
 Zeus, king of the gods and one of the fathers of Orion

The constellation now called Libra (*the Scales*) was not identified as a set of weighing scales by all Greco-Roman sources. In fact, a common interpretation of this constellation was as the claws of Scorpius, the scorpion that killed the hunter Orion. For example, in his catalogue of ancient constellations, the second-century

CLASSICAL MYTHOLOGY OF THE CONSTELLATIONS

astronomer Ptolemy, who wrote in Greek, named them *Chēlai*, "claws," but Latin authors favored the notion of this constellation as a set of scales. How the Romans arrived at this identification is likely traceable to, or linked with, the celestial observers of ancient Babylon, who had interpreted this constellation as scales.

Libra was symbolically linked with the goddesses Astraea and Dikē, both of whom were, at one time or another, identified as the maiden represented by the constellation Virgo. These goddesses were also often confused or conflated with each other. Both were goddesses of justice, and justice—how well or morally people had conducted themselves—was measurable with the Scales of Justice.

Peace and harmony were conditions in which Astraea and Dikē thrived, while violence, greed, and any other sort of evil were their natural enemies. The same well-known myth is told of them both. In days long ago, there was a Golden Age in which people lived harmoniously with each other. There was no need for laws, nor was there commerce, since humans were content to enjoy the earth's bounty, not coveting more than what they had. At this blessed time, gods and mortals also mingled freely with each other. The Golden Age, however, did not last, and was followed by Ages of Silver, Bronze, and finally Iron, each age being more wicked than the next. The gods soon ceased to keep company with humans—all, that is, except Astraea (or Dikē), the goddess of justice. Eventually, in the Iron Age, greed, corruption, and bloodlust reached such a pitch that even the goddess of justice abandoned hope and left the Earth forever, taking up residence instead in the night sky but keeping close watch on mortals below. The Age of Iron is also the age in which we still live today.

If viewed as the claws of Scorpius, the tale associated with this constellation is a very different one. It is a story that centers on Orion, a son of the gods Zeus, Poseidon, and Hermes. How all three gods came to be Orion's fathers is very strange: all three urinated in an animal skin that was subsequently buried. From it a baby of prodigious size was born. This, of course, was Orion. Later, when Orion had matured, he had a number of adventures, some heroic and others quite the opposite. One of them involved the rape of a princess, which left him temporarily blinded. This was the doing of the maiden's angry father. However, once cured of his affliction, Orion lived happily on the island of Crete, keeping company with the goddess Artemis. Together they would walk through the wilderness to hunt. As Orion became ever more skilled at hunting, he also became overly confident, and boasted that he could kill all the animals on the island. At this prospect, Artemis became alarmed, and an equally alarmed Earth goddess, Gaia, sent an enormous scorpion to sting and kill him. Its mission accomplished, the scorpion was set by Zeus among the stars as a memorial to its service to the gods and as a warning

SOUTHERN CELESTIAL QUADRANT III

Libra

to humans. The scorpion was so large, however, that its massive claws became a separate constellation. Orion, for his part, was pardoned and set among the stars, too, but it became his fate always to be fleeing the creature that caused his death. When Scorpius and its claws are seen to rise in the sky, a fleeing Orion sets.

Those who viewed the constellation as the scorpion's claws imagined them as an extension of, or addition to, Scorpius. As scales, Libra is a natural partner to Virgo.

Best visibility: April to June (60°N–90°S). Libra is one of the twelve constellations of the Zodiac, with the Sun passing through it from the end of October to late November.

See also the constellations Orion, Scorpius, and Virgo.

LUPUS

The Wolf

Taking pity upon Chiron, Zeus put him among the constellations with an animal that he holds above the altar for sacrifice.

Hyginus, *Astronomica* 2.38.2

MAIN CHARACTERS:
Amulius, the unlawful king of Alba Longa
Arcas, the son of Callisto by Zeus
Callisto, a maiden pursued by Zeus and transformed by the goddess Hera into a bear
Hermes, the messenger god
Lycaon, an evil king of Arcadia
Mars, the Roman god of war
Numitor, the rightful king of Alba Longa
Remus, twin brother of Romulus
Rhea Silvia, daughter of Numitor and mother of Romulus and Remus
Romulus, twin brother of Remus and the legendary founder of Rome
Vesta, Roman goddess of the hearth
Zeus, king of the gods and father of Arcas as well as Romulus and Remus

SOUTHERN CELESTIAL QUADRANT III
Lupus

CLASSICAL MYTHOLOGY OF THE CONSTELLATIONS

The constellation Lupus (*the Wolf*) is an interesting case of evolving identity. The Greeks and Romans typically referred to this constellation simply as "the beast," without naming any particular species of animal. That generic animal was believed to have been intended for sacrifice on the starry altar (Ara) near the constellation Centaurus. It was only after Ptolemy's *Almagest* with its catalogue of forty-eight constellations was translated into Latin from Arabic in the Middle Ages that the name Lupus ("wolf") was assigned to it. Consequently, there is no particular wolf story that is linked with the constellation.

This being said, there are several classical myths that feature wolves, and over time, these have become associated with the constellation Lupus. The best-known Greek wolf story is that of Lycaon, a legendary king of Arcadia, a region in the mountains of the central Peloponnese. Lycaon was transformed into a wolf as it was that animal, Zeus believed, that best incarnated his true, savage nature. Zeus knew this firsthand. Having learned that humans had been behaving more and more badly, disrespecting each other and the gods, Zeus and the god Hermes set out to test them. Together they went to visit Lycaon, who had earned a reputation for being especially evil. The gods were in disguise, and Lycaon refused to believe Zeus when he revealed that he was actually a god. Deciding to test Zeus's alleged divinity, he served up the god's own grandson Arcas, chopped up and cooked in a stew. Zeus, of course, had no need to taste the stew to know the truth of its ingredients. Arcas was instantly restored to life, ultimately becoming the constellation Ursa Minor (*the Lesser Bear*) and set next to his mother, Callisto, who had also become a constellation (Ursa Major). Lycaon, on the other hand, was transformed into a wolf who went running off into the hills.

The other famous wolf was that which rescued Romulus and Remus, ancient Rome's most famous twins. Romulus and Remus were sons of Mars, the Roman equivalent of Greek war-god Ares, and Rhea Silvia, princess of the Italian town Alba Longa. At the time, Rhea Silvia's uncle, Amulius, held the throne of Alba Longa, having seized the kingship from Numitor, the rightful king. In order to ensure his continued hold on the kingdom, Amulius killed Rhea Silvia's brothers and made her a priestess of Vesta, the goddess of the hearth. Priestesses of Vesta, who were called Vestal Virgins, were required to remain virgins. Failure to do so was punishable by a gruesome death: burial alive. In light of the consequences of breaking the Vestals' vow of chastity, Amulius was most surprised to learn that Rhea Silvia was pregnant and still more surprised when she claimed that Mars was the father. Unwilling now to shed more blood, especially since a god was involved, Amulius imprisoned Rhea Silvia and ordered her newborn twin sons to be placed in a basket and left afloat on the Tiber River. Nature, he believed, would take its course, and the twins would die a bloodless death. However, a she-wolf spotted the twins when the basket washed up on the Tiber's banks and brought them to a cave, where she suckled them. In time, Romulus and Remus were adopted by one of Amulius's shepherds and, after reaching maturity, were able to kill Amulius and restore their grandfather Numitor to the throne.

Romulus and Remus then decided to found their own city. It would be named after Romulus, who killed his brother in a dispute and became Rome's first king.

Best visibility: May to June (34°N–90°S)

See also the constellations Ara, Centaurus, Ursa Major, and Ursa Minor.

OPHIUCHUS

The Serpent-Holder

When all the world rests deep in tranquil sleep, the youth blasted
by his grandfather's fiery bolts rises above the horizon, arms
outstretched, each coiled with a snake.

———

Ovid, *Fasti* 6.734–736

MAIN CHARACTERS:
   Admetus, a legendary king of Thessaly
   Apollo, god of medicine, music, and archery
   Asclepius, a son of Apollo and a god of healing
   Chiron, a noble and highly cultured centaur
   Coronis, the mother of Asclepius
   Cyclopes, one-eyed giants who made Zeus's thunderbolts
   Laocoön, a Trojan priest who realized the Trojan Horse was a trick
   Zeus, king of the Greek gods

Ophiuchus, *the Serpent-Holder* (from *ophis* ["serpent"] + *ochos* ["holder"] in Greek), is usually identified as Asclepius, the Greek god of healing and the patron god of physicians. His signature staff, entwined by serpents, remains a familiar symbol of healthcare today.

Asclepius had not always been a god. Rather, he was the semi-divine and thus mortal son of Coronis, princess of Orchomenos in the Peloponnese, and the god

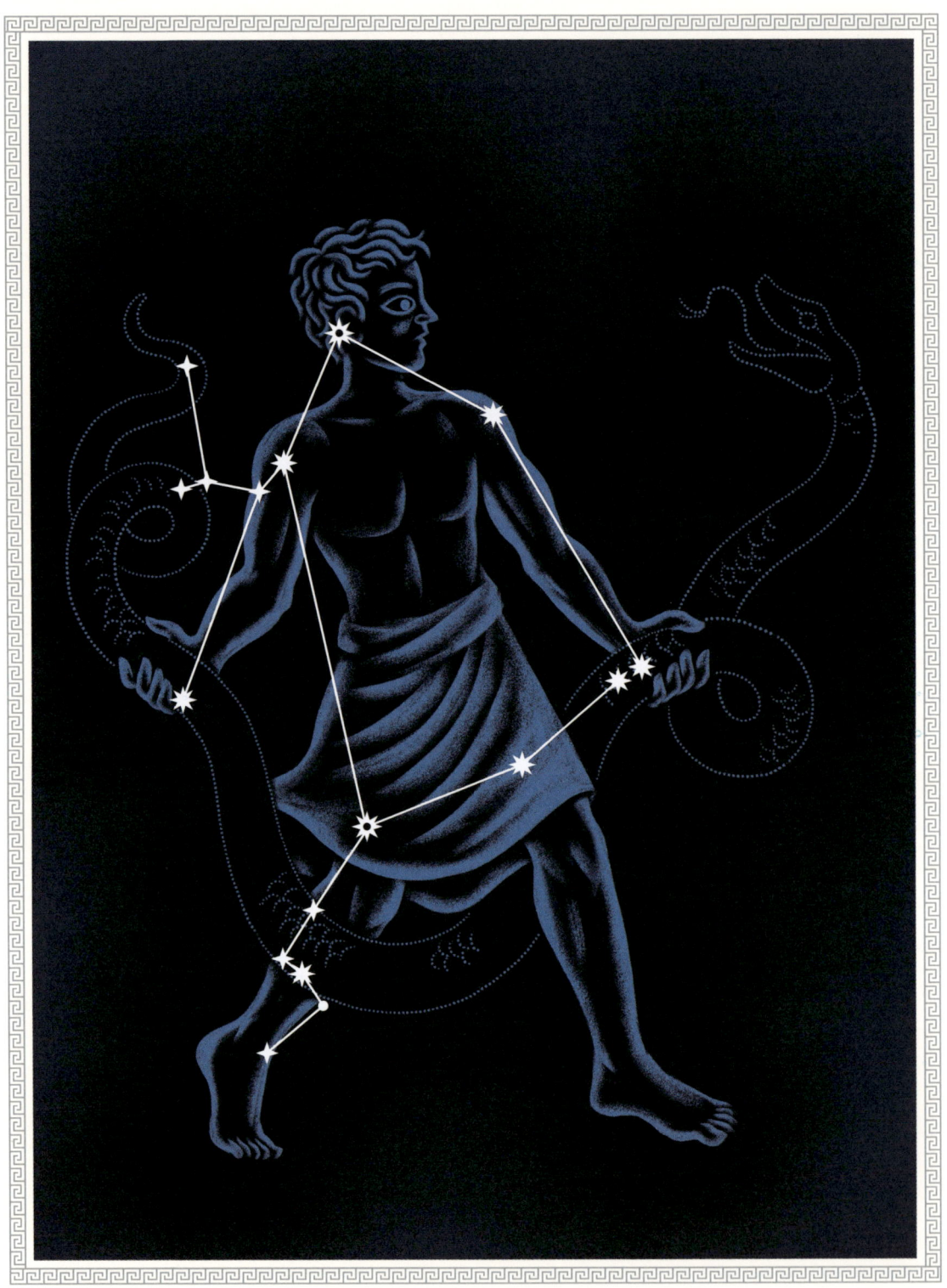

SOUTHERN CELESTIAL QUADRANT III
Ophiuchus

SOUTHERN CELESTIAL QUADRANT III

Apollo. Accounts of his birth vary, but several record the infidelity of Coronis and her punishment by death at the hands of a jealous Apollo. Unbeknownst to Apollo, Coronis was pregnant when he discovered that she had been unfaithful to him and, overcome by anger, shot her with his bow and arrow. The child's father, however, was the god himself. Upon realizing this, Apollo snatched the unborn infant from his mother's body just in time, before the funeral pyre's flames consumed her. The baby Asclepius was delivered to the cave of Chiron, a noble and learned centaur who would later be commemorated by the constellation Centaurus. It was from Chiron that Asclepius learned the art of healing that would become his vocation. It was also an art at which he excelled, so much so that he became discontented with merely curing the ill and set his sights upon reviving the dead, which he eventually achieved. Because Asclepius could now make mortals immortal and, thus, essentially indistinguishable from the gods, Zeus decided to stop him and struck him dead with one of his thunderbolts.

Asclepius's death angered his father, Apollo, who avenged himself for the death of his son by killing the makers of Zeus's thunderbolts, the one-eyed Cyclopes. In this instance, Apollo's weapon was again his bow and arrow, which is commemorated by the constellation Sagitta. Apollo's actions caused Zeus to punish him in turn, and he was made a servant to the Thessalian king Admetus for the period of a year. Zeus did eventually forgive Asclepius's prideful sin, making him divine and lifting him to the heavens. After all, Asclepius was not only the child of a god but Zeus's own grandson. The coiling snake that he holds fast in his hands is the constellation Serpens. Asclepius's mother, too, was memorialized by a constellation, Corvus.

Among other mythological characters sometimes identified as the Serpent-Holder is Laocoön, who played a pivotal, tragic role in the fall of Troy. Laocoön was the Trojan priest who warned his fellow Trojans to beware of the gift—the warrior-filled Trojan Horse—left for them by the Greeks. No sooner had Laocoön uttered his words of warning than a set of serpents emerged from the sea to strangle him and his sons. The Trojans misunderstood this event as reflecting the blasphemous untruth of Laocoön's words. They were mistaken. The Trojan Horse proved to be classical mythology's most famous and lethal trap. The unsuspecting Trojans brought the horse inside the city's walls, and in the dead of night, the Greeks emerged from its wooden belly to set fire to the city, enslave women and their children, and kill any remaining adult men.

In the case of Laocoön as the constellation Ophiuchus, the starry serpent with which he struggles in the heavens represents the serpent that caused his death.

CLASSICAL MYTHOLOGY OF THE CONSTELLATIONS

Asclepius's serpent, on the other hand, taught him much about herbal medicine. In fact, it was from a serpent that Asclepius learned how to revive the dead.

Although the Serpent-Holder's legs and torso reside in Southern Quadrant 3, this large constellation spans the Celestial Equator, the figure's head and shoulders extending into Northern Quadrants 3 and 4.

Best visibility: June to July (59°N–75°S)

See also the constellations Centaurus, Corvus, Sagitta, and Serpens.

SCORPIUS

The Scorpion

Then the goddess roused up another beast, splitting open the
island's hills—a Scorpion, which proving stronger than strong
Orion, wounded him and caused his death. This was because he
had angered Artemis. For this reason, too, people say, that
when the Scorpion rises in the East, Orion takes
flight at the Western verge.

Aratus, *Phaenomena* 641–646

MAIN CHARACTERS:
    Artemis, the goddess of the hunt and protectress of wild animals
    Gaia, the Earth goddess
    Orion, one of classical mythology's most skilled hunters
    Zeus, king of the gods

Scorpius, the scorpion constellation, is intimately linked with the saga of Orion. When the great hunter Orion boasted that he could kill every living thing, in fear—and anger—that what Orion boasted would come to pass, the goddess Artemis took action. Once Orion's companion in the hunt, sharing mutual affection, Artemis now became his nemesis. She, or by some accounts the Earth goddess Gaia, summoned help from the Earth's depths. As the ground split open, a giant scorpion emerged and killed Orion with its fatal sting.

Zeus raised the scorpion to the heavens as a starry, cautionary reminder for humans to exercise restraint in killing other living things. Every life is precious,

SOUTHERN CELESTIAL QUADRANT III
Scorpius

not to be taken for granted. The scorpion's massive claws became a neighboring constellation of their own, the grouping of stars later known as Libra (*the Scales*), or so some in antiquity believed. As for Orion, the goddess Artemis asked that he be elevated to the heavens, too, as a memorial to the bravery he showed at many points in his life. Zeus agreed to this, but on one condition: Orion would flee the scorpion eternally. So it happened that as the constellation Scorpius rises in the heavens, Orion is seen to set.

Best visibility: June to July (44°N–90°S). Scorpius is one of the twelve constellations of the Zodiac, with the Sun passing through it in late November.

See also the constellations Libra and Orion.

SERPENS

The Snake

When Asclepius was ordered to bring young Glaucus back to life and was imprisoned with his lifeless body in a cave, a snake approached and wound its way up Asclepius's staff. In a sudden panic, he killed the snake, striking it again and again with his staff. But then another snake approached carrying an herb in its mouth. This it placed on the head of its slain brother, and— miraculously—the two slithered hastily from the cave.

―――

Hyginus, *Astronomica* 2.14.7

MAIN CHARACTERS:

Asclepius, a son of Apollo who became the Greek god of healing
Glaucus, a young son of King Minos of Crete
Minos, a legendary king of Crete married to Pasiphaë
Minotaur, a son of Pasiphaë who was half-human and half-bull

The snake most often identified with the constellation Serpens (*the Serpent*) is that which famously coiled itself around the staff of the healing god Asclepius. While Asclepius was still a mortal and not yet divine, he had become an expert healer. Not happy simply with healing the sick, he set his sights on bringing the dead back to life. On one occasion, he was tasked with reviving young Glaucus, son of Minos,

SOUTHERN CELESTIAL QUADRANT III
Serpens

the legendary king of Crete who is best known in mythology as the stepfather of the Minotaur. Half-human and half-bull, the Minotaur is associated with the constellation Taurus (*the Bull*).

As for Glaucus, he had met an untimely end by accidentally drowning in a large vat of honey while chasing a mouse. His distraught father commanded the famous physician Asclepius to revive his son, and to ensure that Asclepius did everything in his power to make this happen, he imprisoned him in a cave with Glaucus's body. Asclepius would be released only if and when Glaucus was restored to life.

As Asclepius sat lost in thought, pondering what exactly to do, a snake slithered up his staff. In alarm, Asclepius grabbed the staff and struck the serpent dead. A second snake then appeared carrying an herb in its mouth. Making its way to the dead one, this snake gently placed the herb on its head. No sooner had it done this than both departed as silently and mysteriously as they had come. Amazed at what he had seen, Asclepius applied this herb to Glaucus, who, like the once-dead snake, opened his eyes and breathed again.

Asclepius, like the snake, would be memorialized as a constellation. In the heavens, Asclepius holds the snake fast in his outstretched arms. Unusually, the constellation Serpens consists of two distinct groups of stars that in turn define two distinct parts of the constellation: Serpens Caput (*Serpent's Head*) and Serpens Cauda (*Serpent's Tail*). Asclepius holds the head in his right hand and the tail in his left. Both parts of this constellation extend beyond Southern Quadrant 3, Serpens Caput into Northern Quadrant 3 and Serpens Cauda through Southern Quadrant 4 into Northern Quadrant 4.

Best visibility: June to August (74°N–64°S)

See also the constellations Ophiuchus and Taurus.

SOUTHERN CELESTIAL QUADRANT III

VIRGO

The Virgin

People lived by theft and plunder. No longer were guests safe from their hosts, nor fathers-in-law from their daughters' spouses. Even brothers could not live in peace. Husbands are eager for their wives to die and wives wish death upon their husbands. Evil stepmothers mix dreadful poisons, and sons greedily count their fathers' years. Piety is dead. Now, last of all the gods, the virgin Astraea leaves the blood-drenched earth.

Ovid, *Metamorphoses* 1.144–150

MAIN CHARACTERS:
Astraea, a Greek goddess of justice
Astraeus, a Greek god of prophecy, the stars, and the planets
Demeter, Greek goddess of grain and the harvest
Dikē, a Greek goddess of justice often conflated with Astraea
Dionysus, the democratic god of wine and vegetation
Eos, Greek goddess of the dawn
Erigone, the devoted daughter of the farmer Icarius
Fortuna, the Roman goddess of fortune and equivalent to the Greek goddess Tyche
Icarius, an Athenian farmer who welcomed Dionysus
Themis, the Greek goddess ensuring correct and moral behavior
Tyche, the Greek goddess of good and bad fortune
Zeus, king of the Greek gods

There was little agreement among authors of Greco-Roman antiquity about the identity of the maiden represented by the large constellation Virgo (*the Virgin*). At least half a dozen young women, or their images, were said to have been placed among the stars, but the best known of them were Astraea, a goddess of justice; Tyche, goddess of fortune; Erigone, the devoted daughter of Icarius and a special devotee of the god Dionysus; and Demeter, goddess of grain and the harvest.

Astraea, whose name means "starry maiden," was also known as (or conflated with) the goddess Dikē, both being goddesses of justice. Her parents were

SOUTHERN CELESTIAL QUADRANT III
Virgo

believed to have been either Zeus and Themis, a goddess personifying decent and orderly behavior in communities, or the Titan god Astraeus and Eos, goddess of the dawn. Astraeus, whose name means "starry one," was a somewhat obscure god of prophecy, stars, and the planets.

As it happens, Astraea is best known for her departure from Earth and joining the stars in the heavens. Once upon a time, during the Golden Age when people lived in peace and harmony, knowing no greed and enjoying the earth's bounty, the gods, including Astraea, mingled freely among people. In the following age, the Age of Silver, humans became selfish, violent, disdainful of the gods, and shorter-lived. In the Bronze Age, people became more warlike. Finally, during the Iron Age, a time of outright wickedness, a despondent Astraea abandoned the Earth altogether, the last of all the gods to do so, taking up residence among the stars as a constellation.

Tyche, known to the Romans as Fortuna, has no mythology per se, and it is uncertain how some believed she came to be situated or depicted among the stars. Tyche could be a fickle goddess and could bring either good or bad fortune, wealth or poverty, success or failure.

As for Erigone, she was said to have been depicted among the stars by Zeus out of pity for her terrible fate, a fate unbefitting so devoted and pious a young woman. Erigone and her father Icarius were notable for welcoming the outlandish, effeminate god Dionysus when he came to Greece from the Near East. Dionysus came bringing precious gifts: knowledge of planting grapevines, harvesting grapes, and making wine. All these he taught Icarius and Erigone. Recognizing the potential value of these gifts to humankind, Icarius set out in his oxcart to share his precious, newly made drink. Wine would become the ancient world's staple beverage, drunk diluted with water and regarded as the liquid essence of the liberating, democratic god Dionysus himself.

Icarius first came upon some shepherds, who drank freely but, instead of feeling gratitude, became drunk and paranoid. They killed Icarius and hid his body well. Erigone became worried over her father's lengthy absence and began to look for him. Their faithful dog Maera had been searching, too, and led Erigone to her father's corpse. A despondent Erigone longed for death and hung herself on a nearby tree. Having lost her human family, the equally despondent dog leaped into a well to her death. The selflessness and devotion of all three did not go unnoticed by the gods, and Zeus placed images of them among the stars as constellations: Erigone as Virgo, Icarius as Boötes, and Maera as Canis Minor (or, by some accounts, either the star Canicula or the constellation Canis Major).

The belief in Demeter as the maiden of Virgo is puzzling, as Demeter was thought of as a mature woman and not a young, virginal one. In classical mythology and religion, Demeter was the mother of Persephone, who became the bride of Hades and part-time queen of the Underworld. There is no clear explanation of how it came to pass that a likeness of Demeter was featured among the stars.

Representations of Virgo vary and, interestingly, they reflect attributes of several of the mythological characters associated with this constellation. The maiden of Virgo is generally winged, as Astraea and Tyche were thought to be in antiquity, and she sometimes holds or reaches out to the scales of the constellation Libra with her right hand. Astraea and Tyche used scales to measure the degree of a person's fortune or the length and quality of their life. However, Virgo's maiden is often shown holding a sheaf of wheat in her left hand, and wheat is the symbol of the grain goddess Demeter.

Best visibility: April to June (67°N–75°S). Virgo is one of the twelve constellations of the Zodiac, with the Sun passing through it from mid-September to the end of October. Although Virgo is located largely in Southern Quadrant 3, the constellation does extend into the Northern Celestial Hemisphere: the maiden's head resides in Northern Quadrant 2, her hair flows into Northern Quadrant 3, and a wing, together with a portion of her right leg, spreads to Northern Quadrant 3.

See also the constellations Boötes, Canis Minor, Canis Major, and Libra.

SOUT
CELE
QUAD

HERN
STIAL
RANT
IV

SOUTHERN CELESTIAL QUADRANT IV
Aquarius

SOUTHERN CELESTIAL QUADRANT IV

AQUARIUS

The Water-Bearer

> All the gods of Olympus were sitting with Zeus in his great
> hall, gathered there on gilded thrones. As they feasted, lovely
> Ganymede poured nectar, the gods' fragrant drink,
> making his rounds to fill their goblets.
>
> ―――
>
> Nonnus, *Dionysiaca* 27.241–245

MAIN CHARACTERS:

Athena, the patron goddess of Athens and goddess of defensive war
Cecrops, a snake-man and an early king of Athens
Deucalion, a son of Prometheus known for his piety
Ganymede, a prince of Troy who became Zeus's cupbearer
Poseidon, the god of the sea
Prometheus, a Titan god known for helping humankind
Pyrrha, the pious wife of Deucalion
Themis, a goddess of justice
Zeus, king of the gods

In antiquity, several different mythological characters were identified as the figure represented by the constellation Aquarius (*the Water-Bearer*). The most widespread notion was that this was Ganymede, a young shepherd and, as it happened, a Trojan prince who was extraordinarily handsome. Zeus caught sight of him one day as he tended his flocks on Mount Ida, southeast of Troy, and fell in love with him. By some accounts, Zeus sent his sacred bird, an eagle, to snatch up Ganymede and carry him to Mount Olympus. By others, it was Zeus himself who did this, having transformed himself into an eagle. On Mount Olympus, Zeus made Ganymede divine, now visible to mortals as a constellation, and gave him the privileged job of serving as cupbearer of the gods, ensuring that their cups were always full. The eagle (Aquila) that had carried off Ganymede was also memorialized among the stars and resides adjacent to the youth.

Another character identified as Aquarius is Deucalion, a son of the Titan god Prometheus. Deucalion was a king of Phthia in Thessaly, northern Greece, and he was married to a woman named Pyrrha. At a time when wickedness ran rampant among humans, Deucalion and Pyrrha were notable for their piety, and for this reason they alone were spared when Zeus decided to eliminate humankind. Zeus briefly considered setting the Earth aflame with his thunderbolts but became concerned that the heavens, too, would be impacted. He decided instead to resort to causing a great flood and enlisted the aid of his brother Poseidon. It rained and stormed for nine whole days. All villages and fields were destroyed. People and their flocks alike were swept away and drowned. Those living things that did not drown later died by starvation. Only Deucalion and Pyrrha escaped with their lives, by weathering the storm in a small boat. When the rains abated, this boat washed up on the slopes of Mount Parnassus. While grateful to be alive, Deucalion and Pyrrha were grief-stricken at the loss of all other human life. So, they made their way to a nearby temple of Themis, a goddess of justice, and asked her how they might restore the human race. Through an oracle, the goddess told them that they should throw the bones of their mother over their shoulders. Pyrrha believed this to be sacrilege—who would desecrate the remains of a parent?—and would not do it. Deucalion, however, interpreted the oracle less literally and understood the bones of their mother to be stones and their mother to be the mother of all living beings, Mother Earth. Now both cast stones over their shoulders and saw these change their shapes into the familiar forms of men and women. Other creatures now rose spontaneously from the earth as well.

Cecrops, a legendary king of Athens—by some accounts the first—is yet another figure considered to be depicted by the constellation Aquarius. Cecrops, a snake-man, was born from the earth, and in place of human legs, he had a serpent's tail. Cecrops was credited with a number of cultural advancements. He had temples built in the territory of Athens and established the worship there of Athena and of Zeus. He taught people how to write and how to properly bury their dead. He also abolished the practice of human sacrifice, substituting drink offerings, the pouring of liquid sacrifices (libations), in its place. He was also the individual credited by some with having decided the outcome of the contest between Athena and Poseidon for the patronage of Athens. Athena caused an olive tree to spring up on the Acropolis as a gift to the Athenian people, and Poseidon produced a horse (or, according to some authors, a saltwater spring symbolizing naval power) by striking a rock with his trident. Cecrops judged the olive tree, which in fact became the mainstay of the Athenian economy, to be the more valuable gift, and Athena became patron goddess of Athens.

This constellation depicts a man pouring water from a hefty vessel. If it is Ganymede, then he is shown pouring water for the gods to drink. If it is

Deucalion, his vessel contains the abundant floodwaters that he survived, and if it is Cecrops, he is shown pouring an offering of libations to the gods.

Best visibility: August to October (65°N–86°S). Aquarius is one of the twelve constellations of the Zodiac, with the Sun passing through it from mid-February to mid-March.

See also the constellation Aquila.

CAPRICORNUS

The Sea-Goat

People say that Pan leaped down into the flowing waters, making the lower part of his body like that of a fish, and the rest a goat, in this way escaping the clutches of Typhon. In admiration of his cleverness, Zeus placed his likeness among the stars.

———

Hyginus, *Astronomica* 2.28.1

MAIN CHARACTERS:
Aphrodite, Greek goddess of love and desire
Cupid, the son of Aphrodite by Ares
Hermes, the messenger god
Pan, patron deity of herdsmen, flocks, and fields
Syrinx, a nymph loved by Pan
Typhon, a hundred-headed monster who waged war on the gods of Mount Olympus
Zeus, king of the gods and son of the Titan god Cronus

The constellation Capricornus, which is known in English as *the Sea-Goat* (or simply *the Goat*), is typically represented as a goat with the tail of a fish. This constellation had also been recognized by ancient Babylonian stargazers and called a goat-fish by them as well, so it is highly likely that Babylon was the original source of this constellation's particular hybrid nature. The name Capricornus, however, literally means "goat's horn" or "horned goat" and not "sea-goat."

As far as the mythology of this constellation is concerned, Capricorn was thought of as being a representation of Pan, patron deity of herdsmen, flocks, and

SOUTHERN CELESTIAL QUADRANT IV
Capricornus

fields. While Pan's parentage was debated—he was known as a son of Zeus or of Hermes, the messenger god—ancient authors and artists agreed on his appearance. His head and torso were largely human, but he had the horns, beard, legs, and tail of a goat. His favorite haunts were woodland places where he could be found playing the pipe. That pipe, his signature instrument, was once a beautiful nymph named Syrinx, who feared Pan's amorous pursuit and escaped him by changing into a clump of reeds. Pan noticed that these reeds made a lovely sound when the wind blew through them, so he harvested the reeds and bound them together, in this way inventing the shepherd's pipe.

How Pan came to be depicted among the stars is a story quite similar to that of Aphrodite and Cupid, the deities linked with the constellation Pisces (*the Fishes*). When the fearsome, many-headed monster Typhon made an assault on Zeus and all the other gods, Pan fled and leaped into a stream. In the water, he grew a fish's tail and was able to swim to safety. Zeus so admired the cleverness of his escape that he placed an image of the fish-tailed Pan among the stars.

There was a slightly different version of Pan's fishy transformation, too. The occasion was a battle between the older-generation Titan gods and the Olympians. The Olympians were Zeus, his siblings, and his children who lived on Mount Olympus, and the Titans were Zeus's father and the latter's siblings. What was at stake was rulership of the universe, and the fighting was so fierce that heaven and earth shook. In this battle, Pan was said to have played a critical role. His sudden, unexpected shout caused the frightened Titans, as well as the monster guarding a wounded Zeus, to retreat in a heightened fear—a fear, not coincidentally, called "panic." The constellation memorializing this event took a fishlike form to reflect Pan's hurling shellfish at the Titans to hasten their retreat.

Best visibility: August to September (62°N–90°S). Capricornus is one of the twelve constellations of the Zodiac, with the Sun passing through it from mid-January to mid-February.

See also the constellation Pisces.

CORONA AUSTRALIS

The Southern Crown

Dionysus, arriving at the entrance to the Underworld and preparing to descend, left the crown, a gift from Aphrodite, at that place which now, in memory of that event, is called Stephanos. He didn't want to take that immortal gift with him lest it be polluted by contact with the dead. When he brought his mother back unharmed to reside amongst the living, he is said to have placed the crown in the stars as a lasting memorial to her.

Hyginus, *Astronomica* 2.5.2

MAIN CHARACTERS:
Dionysus, the god of wine, vegetation, and the theater
Hera, queen of the Greek gods and wife of Zeus
Semele, a princess of Thebes and mother of Dionysus
Zeus, king of the Greek gods and father of Dionysus

Unusually, there is no really noteworthy or dominant mythological tradition associated with Corona Australis (*the Southern Crown*), though the constellation was certainly recognized in antiquity. Some believed that it was a wreath that slipped off the head of the centaur depicted in the constellation Sagittarius (*the Archer*). In antiquity, wreaths functioned not only as crowns for royalty or the gods but also as symbols of extraordinary status, thus being given as prizes to victors in the Olympic Games or being worn in ritual or other celebratory contexts. Centaurs were frequently depicted in ancient art as wearing wreaths, so a wreath for the constellation Sagittarius's centaur was wholly appropriate.

Others, however, believed that this constellation was a wreath that the god Dionysus placed in the southern sky to commemorate his mother Semele, her tragic death, and her later resurrection. Semele, a princess of Thebes, was one of many mortal women who were pursued and impregnated by the god Zeus. Hera, Zeus's jealous wife, became aware of her husband's infidelity with Semele and plotted her vengeance. Disguising herself as the family's trusted nursemaid, Hera approached Semele's sisters and told them that Semele was with child—worse, she was with child out of wedlock, and the father could be any scoundrel. Their

SOUTHERN CELESTIAL QUADRANT IV
Corona Australis

curiosity piqued, the sisters questioned Semele about the father. She swore that her lover was none other than the king of the gods, but her sisters' continued probing after the father's true identity cast a seed of doubt. She should ask that her divine lover prove his divinity, they argued. And so, filled with misgivings, Semele asked Zeus for a favor when he next visited her. He promised without asking what the favor was. For a supposedly omniscient god, it was a grievous error. Zeus was the god of oaths and the protector of promises, so he could not break his word when Semele then asked him to reveal himself in his full divinity. With the heaviest of hearts, Zeus revealed himself in the fullness of his divine splendor, and as he knew would happen, Semele burst into flames. The intensity of his pure divine essence was just too much for any mortal to bear.

While Zeus was unable to save Semele, he was able to save the infant Dionysus. Plucking the unborn fetus from Semele's womb, Zeus placed it in his thigh to incubate. In nine months' time, the baby Dionysus miraculously emerged from his father's throbbing thigh. And, in the course of time, Dionysus descended into the Underworld to retrieve his mother from the dead, leaving behind a wreath of myrtle leaves to commemorate this fact. He then further memorialized his mother's ordeal by placing a wreath in her honor among the stars as well.

Best visibility: July to August (44°N–90°S)

See also the constellation Sagittarius.

PISCIS AUSTRINUS

The Southern Fish

It is believed that when the goddess Isis was in labor, the fish
came to her aid, and as a reward for this kindness she placed
the fish and its young, about whom we have
spoken before, among the stars.

Hyginus, *Astronomica* 2.41.1

MAIN CHARACTERS:
Aphrodite, Greek goddess of love and desire
Horus, Egyptian god of the sky and sun
Isis, Egyptian goddess of fertility and abundance

SOUTHERN CELESTIAL QUADRANT IV
Piscis Austrinus

Osiris, Egyptian god of the Underworld and husband of Isis
Seth, the violent Egyptian god of thunderstorms and elemental disorder

Piscis Austrinus (*the Southern Fish*) is also known as Piscis Australis and, as its name suggests, it is located in the Southern Celestial Hemisphere. This fish was thought to be the parent of the two fishes of the constellation Pisces. Interestingly, the explanation of Piscis Austrinus's appearance in the skies is very similar to that of Pisces (*the Fishes*). Pisces, it was believed, commemorated the goddess Aphrodite's rescue by a pair of fishes. In the case of Piscis Austrinus, however, the divine protagonist is a pregnant Isis, an Egyptian mother-goddess who became enormously popular among Greeks and Romans alike. Isis's child was the god Horus, and her mate, Horus's father, was Isis's own brother, the god Osiris. Seth, the violent god of storms, had murdered Osiris and went in pursuit of Isis, too, as he considered her unborn son a threat. In the course of her flight from Seth—her predicament compounded by the sudden onset of labor—Isis leaped into a river and a fish came to her rescue. This, of course, was the fish that Piscis Austrinus commemorates.

Isis did then deliver her child safely while taking refuge in a thick clump of papyrus on the river's banks. She also restored her husband, Osiris, to life after searching all of Egypt for, and reassembling, the parts of his dismembered body.

Piscis Austrinus is pictured swimming upside down and with its mouth open, drinking all the water flowing from the water-bearer Aquarius's cup.

Best visibility: September to October (53°N–90°S)

See also the constellations Aquarius and Pisces.

SOUTHERN CELESTIAL QUADRANT IV

SAGITTARIUS

The Archer

Chiron groaned as he pulled the arrow's iron tip from his flesh, and Hercules sighed in sorrow, as did the young Achilles. Chiron mixed herbs collected from the Pagasean hills, and, to no avail, tried to soothe the wound. The voracious poison overwhelmed the treatments; the pestilence penetrated deep into his bones and his entire body. And so, the blood of Lerna's dreaded Hydra mingled with his own, allowing no time for rescue.

Ovid, *Fasti* 5.399–407

MAIN CHARACTERS:

Achilles, a Greek hero famed for his exploits during the Trojan War
Argonauts, the men who sailed with Jason on the ship *Argo*
Chiron, the gentlest and most cultured of all centaurs
Cronus, a Titan god who fathered Chiron as well as the Olympian gods
Crotus, a son of the god Pan and the nymph Eupheme
Eupheme, a nymph and the mother of Crotus
Hercules, the greatest of all Greek heroes and celebrated for his Twelve Labors
Hydra, a many-headed water snake slain by Hercules
Jason, the hero who led the expedition of Argonauts to retrieve the golden fleece
Muses, patron goddesses of the arts who lived on Mount Helicon
Pan, a rustic god of herds and herdsmen who was himself part goat
Philyra, a nymph and the mother of Chiron
Pholus, a kind and gentle centaur
Rhea, a Titan goddess who was the sister and wife of Cronus
Zeus, king of the gods

The constellation Sagittarius (*the Archer*) was envisioned as a centaur, half-human and half-horse, galloping through the heavens with a bow and arrow in his hands, ready to strike. Although this centaur's appearance, in particular his weapon, was different from that of the constellation Centaurus, the identity and mythology of the two constellations became confused even in antiquity.

SOUTHERN CELESTIAL QUADRANT IV
Sagittarius

SOUTHERN CELESTIAL QUADRANT IV

Unsurprisingly, the centaur most often linked with Sagittarius was Chiron, classical antiquity's most famous and also most gentle and cultured centaur. Chiron was said to be the offspring of the Titan god Cronus and a nymph by the name of Philyra. Cronus was married to his sister, Rhea, and with her became father to Zeus and the other gods of Mount Olympus. So as to act on his improper desire for Philyra without rousing Rhea's suspicion, Zeus changed himself into a stallion in order to lie with her. Much to Philyra's horror, she later bore Chiron, a mixed-breed, illegitimate son.

In spite of his strange parentage, Chiron grew up to be unusually talented. He lived apart from the other centaurs, who had different parents and were unruly, unlawful creatures, and he became highly skilled in a range of arts. He was an expert musician, prophet, hunter, and healer, and for this reason he became the teacher, and foster father of sorts, of a host of Greek heroes, among them Hercules, Achilles, and Jason, leader of the Argonauts. It was Chiron's later great misfortune to be wounded accidentally by one of Hercules's arrows—a wound inflicted on himself when he dropped an arrow on his foot. Hercules's arrows were poisonous, having been dipped in the dreaded many-headed Hydra's venom-laden blood, and Chiron's great skill as a healer was no match for the arrow's potency. As it happened, Chiron was immortal, but his pain was so great that he longed for death. Zeus granted his wish by way of compromise, allowing him to leave his bodily existence behind but re-creating him with stars as a constellation to grace the heavens. Ironically, as a constellation he holds the weapon that ended his Earthly life: a bow and arrow. Zeus would later do the same for Hercules, who joined Chiron and the Hydra in the night sky. These three would become the constellations Hercules, Sagittarius, and Hydra.

Another centaur identified as Sagittarius was the lesser-known centaur Pholus, and essentially the same story of his earthly death and resurrection as a constellation was told of him. He, too, was unusually kind for a centaur and was fatally wounded by his own hand while inspecting one of Hercules's lethal arrows. Turning Pholus into a constellation was Zeus's reward for his being a caring, generous soul.

A third character associated with the constellation Sagittarius is Crotus, a son of the rustic god Pan and the nymph Eupheme, who lived on Mount Helicon and was the nurse of the Muses when they were infants. Pan was part human and part goat in appearance, but Crotus apparently did not inherit his father's hybrid form. Crotus was not a centaur, nor was he part goat. He was fully human in appearance. Not much more is said about Crotus other than that he was a companion of the Muses, a skilled hunter, a fast runner, and proficient in the arts. Upon his death, the Muses asked Zeus to commemorate him with a constellation, which he did. The constellation does not replicate Crotus's actual appearance but rather symbolically represents his various skills and attributes. As a constellation, Crotus was given a horse's body to reflect his running speed and a bow and arrow to reflect his skill at hunting.

Best visibility: July to August (44°N–90°S). Sagittarius is one of the twelve constellations of the Zodiac, with the Sun passing through it from mid-December to mid-January.

See also the constellations Centaurus, Hercules, and Hydra.

CONSTEL
AFTER P

LATIONS
TOLEMY

CLASSICAL MYTHOLOGY OF THE CONSTELLATIONS

There are currently eighty-eight constellations officially recognized by the International Astronomical Union (IAU). These include the forty-eight ancient constellations listed by Ptolemy (Claudius Ptolemaius) in his work the *Almagest*, which was likely written sometime between 100 and 174 CE, as well as forty constellations identified in the millennia after Ptolemy. The newer constellations resulted from ongoing efforts to map the entire night sky, filling in uncharted portions of the Northern and Southern Celestial Hemispheres. All of them, like Ptolemy's constellations, were assigned names in Latin, but tenuous links with classical mythology can only be established for a few of them. All are listed in the following pages.

While many individuals proposed many new constellations, most of these were not subsequently accepted by the scientific community. In the end, new constellations introduced by only a very small number of astronomers withstood the test of time, earning official acceptance. The astronomers in question are Johannes Hevelius, Nicolas-Louis de Lacaille, Petrus Plancius (working both alone and with Pieter Dirkszoon Keyser and Frederick de Houtman), and Caspar Vopel.

Petrus Plancius, a Dutch cartographer, led the way in charting the Southern Celestial Hemisphere, but his was, at least in part, a collaborative effort. Although sixteen modern constellations are attributed to him, twelve of them were based on the observations of navigators Pieter Dirkszoon Keyser and Frederick de Houtman, whom Plancius had instructed to map the southern sky in the course of an expedition to the East Indies. Published by Plancius (depicted on a celestial globe) in 1597/1598, these represent animals considered to be exotic by the European expeditions that came across them.

Relying on naked-eye observations, Polish astronomer Johannes Hevelius (1611–1687) identified seven of the post-Ptolemaic IAU constellations, publishing them in his 1687 constellation atlas entitled *Firmamentum Sobiescianum sive Uranographia* ("Sobieski's Heaven, or Star Catalogue," Sobieski being a reference to the Polish king Jan III Sobieski).

The French astronomer Nicolas-Louis de Lacaille (1713–1762) contributed seventeen constellations, fourteen of them located in the Southern Celestial Hemisphere. De Lacaille's constellations were based on observations made from an observatory that he built at the Cape of Good Hope, in what is now South Africa. De Lacaille named almost all of his constellations after new, revolutionary scientific instruments invented during the Enlightenment, among them Microscopium (*the Microscope*) and Telescopium (*the Telescope*). Others were named after tools and equipment used in the arts, with one notable exception, Mensa (*the Table Mountain*), which was named after a distinctive feature of the South African landscape.

Unlike Plancius, Hevelius, and de Lacaille, German cartographer and instrument-maker Caspar Vopel (1511–1561) is recognized not for introducing any new, accepted constellations but, instead, for promoting acceptance of a particular asterism (a distinct but "informal" grouping of stars), already recognized in antiquity, as a bona fide constellation. That ancient asterism was Coma Berenices (*the Hair of Berenice*).

As a final note, those studying the IAU list may notice that only forty-seven, not forty-eight, constellations are shown as attributed to Ptolemy. This is because Ptolemy's enormous constellation Argo Navis (*the Ship* Argo) was considered to be too large by later astronomers and, accordingly, was divided into three new constellations, each of them representing a part of the original ship: Carina (*the Keel*), Vela (*the Sails*), and Puppis (*the Stern*). These three modern constellations now replace Ptolemy's Argo Navis. The astronomer responsible for this division was de Lacaille.

CLASSICAL MYTHOLOGY OF THE CONSTELLATIONS

PLANCIUS'S CONSTELLATIONS

CAMELOPARDALIS

The Giraffe

The constellation Camelopardalis, which represents a giraffe, was introduced by Plancius in the year 1612. The Latin name of this constellation is a combination of the ancient Greek words for camel (*kamēlos*) and leopard (*leopardos*), meaning "camel-leopard." The name reflects the fact that giraffes have long necks like camels and are spotted like leopards.

This faint but large constellation is located in the Northern Celestial Hemisphere and spans Northern Quadrants 1, 2, and 3. It lies between the ancient constellations Ursa Major (*the Greater Bear*) and Cassiopeia.

Best visibility: December to May (90°N–3°S)

COLUMBA

The Dove

The constellation Columba, its name meaning "the dove" in Latin, was depicted by Plancius on a celestial map and globe in the late sixteenth century. Columba is a shortened form of the original name Columba Nohae, "Noah's Dove," and represents the dove that Noah was said to have released from his famous ark to determine if the floodwaters had receded and there was land nearby. The dove returned with an olive branch in its beak, a clear indicator of land, and Columba is shown flying through the heavens with this branch in its beak.

The constellation Columba is located in the Southern Celestial Hemisphere and spans Southern Quadrants 1 and 2. The dove flies near the ancient constellation Canis Major (*the Greater Dog*) and behind the ship of Argo Navis (*the Ship* Argo), which Plancius re-envisioned as Noah's ark.

Best visibility: January (46°N–90°S)

CRUX

The Southern Cross

The small but bright and distinctive constellation Crux, which means "the cross" in Latin, was known in antiquity but was considered part of the ancient constellation Centaurus (*the Centaur*). Precisely who should be credited with introducing this group of stars as its own constellation is unclear, though it was certainly known as such in the sixteenth century. An early illustration of the constellation as a cross was done by Plancius.

Crux, the smallest of all the eighty-eight IAU constellations, is located in Southern Quadrant 3, where it is surrounded by the ancient constellation Centaurus (*the Centaur*) as well as the modern constellations Musca (*the Fly*) and Carina (*the Keel*), which had been part of the enormous ancient constellation Argo Navis (*the Ship* Argo).

Best visibility: April to May (25°N–90°S)

MONOCEROS

The Unicorn

The constellation Monoceros (*the Unicorn*) has a name derived from ancient Greek but represents a mythical creature unknown in the classical world. Like the interestingly named animal constellation Camelopardalis (*the Giraffe*, literally "camel-leopard"), Monoceros was introduced by Plancius in the year 1612 when he included it on an illustrated globe.

Monoceros lies on the Celestial Equator, spanning Northern Quadrant 2 as well as Southern Quadrant 2. The unicorn gallops through the heavens surrounded by a host of ancient constellations: Gemini (*the Twins*), Orion, Lepus (*the Hare*), Canis Major (*the Greater Dog*), Hydra (*the Water Snake*), and Canis Minor (*the Lesser Dog*). Puppis, the "stern" of the ancient mega-constellation Argo Navis (*the Ship* Argo), lies nearby as well.

Best visibility: January to February (78°N–78°S)

PLANCIUS'S COLLABORATIVE CONSTELLATIONS

APUS

The Bird of Paradise

The story of this constellation's name is a tangled one. Apus is a shortened form of Apus Paradiseus in Latin, but Plancius originally called it Paradysvogel Apis Indicus, a combination of Dutch, ancient Greek, and Latin that literally means "the paradise-bird bee of India." *Apis*, the Greek word for "bee," however, may have been a misprint for *avis*, the Latin word for "bird," as the constellation was always visualized as a bird. In any event, the constellation subsequently became variously known as Avis Indica, Apus Indica, and finally, Apus.

Apus, in turn, is a Latinized version of the Greek word *a-pous*, which means "lacking feet." Interestingly, it appears that specimens of the spectacularly plumed New Guinean bird of paradise introduced to Westerners in the sixteenth century had had their feet removed for "aesthetic purposes" by local traders.

The constellation Apus represents the New Guinean bird of paradise in flight. It is located in Southern Quadrant 3, where it is surrounded by the ancient constellation Ara (*the Altar*) and Centaurus (*the Centaur*) as well as the modern constellations Triangulum Australe (*the Southern Triangle*), Circinus (*the Drawing Compass*), Musca (*the Fly*), Chamaeleon (*the Chameleon*), Octans (*the Octant*), and Pavo (*the Peacock*).

Best visibility: May to July (7°N–90°S)

CHAMAELEON

The Chameleon

Chamaeleon is a Latinized form of the ancient Greek *khamaileōn*, a compound of *khamai* ("on the ground") and *leōn* ("lion"), which is translated as "chameleon" but literally means "ground-lion." In other words, the name of this distinctive species of lizard is based on the fact that it creeps along the ground on short legs rather than the fact that it changes color to camouflage itself.

Originally, the small constellation Chamaeleon was represented as a chameleon stretching its tongue toward a fly overhead. The fly, its prey, is the constellation now known as Musca (*the Fly*). Chamaeleon lies in Southern Quadrant 2 but extends somewhat into Southern Quadrant 3. The same is true of Musca. Close to the Southern Celestial Pole, it is surrounded by the modern constellations Volans (*the Flying Fish*), Mensa (*the Table Mountain*), Octans (*the Octant*), and Apus (*the Bird of Paradise*). Vela (*the Sails*), a part of the ancient constellation Argo Navis (*the Ship Argo*) is nearby.

Best visibility: February to May (7°N–90°S)

DORADO

The Golden Fish

Unusually, Dorado, this constellation's name, is not a Latin one. Rather, it is the Spanish word for mahi-mahi (the dolphinfish, *Coryphaena hippurus*), literally meaning "golden." The tropical (and subtropical) fish dorado is distinguished by its colors, which are a brilliant combination of gold, green, and blue.

The small constellation Dorado represents a dorado swimming through the heavens with its favorite meal, a flying fish, leaping overhead. The flying fish, too, is depicted by a constellation: Volans (*the Flying Fish*).

The constellation Dorado is located in Southern Quadrant 1, where it is surrounded by the modern constellations Pictor (*the Painter's Easel*), Reticulum (*the Reticle*), Hydrus (*the Male Water Snake*), Mensa (*the Table Mountain*), and, of course, Volans.

Best visibility: December to January (20°N–90°S)

GRUS

The Crane

Grus means "the crane" in Latin. There are fifteen species of crane, which are large, distinctive, long-legged birds. It is not known which species was originally envisioned as gracing the heavens, but the Southeast Asian sarus crane (*Antigone antigone*), the tallest of the cranes at up to nearly 2 meters (6 feet) in height, has been suggested as a likely candidate.

The constellation Grus, which had been regarded as part of the constellation Piscis Austrinus (*the Southern Fish*) in antiquity, represents a crane in flight with fully outstretched wings. It is located in Southern Quadrant 4, where it forms part of a group of other bird constellations that include Pavo (*the Peacock*), Phoenix (*the Phoenix*), and Tucana (*the Toucan*).

Best visibility: September to October (33°N–90°S)

HYDRUS

The Male Water Snake

The name of the constellation Hydrus means "the water snake" in Latin. In an effort to distinguish it from the larger, ancient constellation Hydra, the many-headed female snake slain by Hercules, this snake (Hydrus) is specifically male, which is indicated by the "-us" ending of its name.

The small constellation Hydrus is located in Southern Quadrant 1 and represents a water snake winding its way through the southern sky. Around it are the modern bird-constellations Phoenix (*the Phoenix*) and Tucana (*the Toucan*) as well as Mensa (*the Table Mountain*), Reticulum (*the Reticle*), and Horologium (*the Pendulum Clock*).

Best visibility: October to December (8°N–90°S)

INDUS

The Indian

While most of Plancius's collaborative constellations represent exotic animals, Indus (*the Indian*) is a clear exception. It is the only one of these to represent a person. While one can say with certainty that the constellation is meant to represent an indigenous person, it is not known what country's indigenous population—or maybe even a specific person that the Dutchmen met in the course of their travels—the constellation depicts.

The constellation Indus is depicted as a male figure wearing a loincloth and holding a spear (and sometimes multiple spears or arrows) and lies in Southern Quadrant 4 between the modern constellations Pavo (*the Peacock*) and Tucana (*the Toucan*).

Best visibility: August to October (15°N–90°S)

MUSCA

The Fly

Flies are certainly not exotic, but they are some exotic creatures' standard fare. The starry lizard represented by the constellation Chamaeleon (*the Chameleon*) is such an exotic creature, and it was grouped with Musca (*the Fly*) in the night sky. The chameleon is shown stretching its tongue toward the fly hovering just overhead.

Like its thematic partner Chamaeleon, the constellation Musca lies in Southern Quadrants 2 and 3, flying between Chamaeleon and the bright, distinctive modern constellation Crux (*the Southern Cross*).

Best visibility: April to May (14°N–90°S)

PAVO

The Peacock

If the constellation Pavo (*the Peacock*) represents a species that was exotic in the eyes of European explorers, then it is likely the green Southeast Asian peacock (*Pavo muticus*) rather than the familiar blue peacock (*Pavo cristatus*) that was imagined as gracing the southern sky.

Pavo was originally shown as having a very luxurious, outspread tail, but its tail feathers were shortened by the later French astronomer Nicolas-Louis de Lacaille to make space for his new constellation Telescopium (*the Telescope*).

Pavo's peacock struts through the heavens of Southern Quadrant 4, where it is surrounded by the ancient constellation Ara (*the Altar*) as well as the modern constellations Triangulum Australe (*the Southern Triangle*), Apus (*the Bird of Paradise*), Octans (*the Octant*), Tucana (*the Toucan*), Indus (*the Indian*), and, of course, Telescopium.

Best visibility: July to September (15°N–90°S)

PHOENIX

The Phoenix

Phoenix is the largest of Plancius's twelve collaborative constellations and represents the mythological phoenix bird that, in classical antiquity, was believed to live in the wilderness of Arabia. It was described as a large bird of great beauty, with red, gold, blue, and purple feathers, and was credited with having a lifespan of more than half a millennium. This bird, it was said, made its own funeral pyre scented with aromatic desert gums and spices (frankincense, myrrh, and cinnamon), and after its body had turned to ash, a new young bird rose from it. In this way, the phoenix was reborn to live again.

The constellation Phoenix is located in Southern Quadrants 1 and 4 at the southernmost end of the ancient constellation Eridanus's (*the River*) celestial stream. Appropriately, the mythical phoenix is accompanied in the night sky by other starry birds: Grus (*the Crane*) and Tucana (*the Toucan*).

Best visibility: October to November (32°N–90°S)

TRIANGULUM AUSTRALE

The Southern Triangle

Triangulum Australe, "the southern triangle" in Latin, represents an instrument called a triangle that is triangular in shape and is used for technical drawing.

The small but distinct constellation Triangulum Australe is visually defined by three bright stars marking the points of an (almost) equilateral triangle. It lies in Southern Quadrant 3 in the Milky Way and is grouped with other modern constellations representing instruments used to survey and map the heavens: Circinus (*the Drawing Compass*) and Norma (*the Set Square*). Other constellations around it are the ancient Ara (*the Altar*) and Centaurus (*the Centaur*), as well as the modern groupings Apus (*the Bird of Paradise*) and Pavo (*the Peacock*).

Best visibility: June to July (19°N–90°S)

TUCANA

The Toucan

Tucana, Latin for "the toucan," is likely not a bird seen by Keyser and de Houtman on their East Indies expedition. Toucans are not native to the East Indies but rather to Central and South America, which is where this species is believed to have been sighted by Keyser in the course of another, earlier expedition.

The constellation Tucana represents a toucan with a sprig of berries in its beak. This distinctive bird—notable for its large bill—rests in Southern Quadrants 1 and 4, where it is surrounded by the bird constellations Phoenix (*the Phoenix*), Grus (*the Crane*), and Pavo (*the Peacock*). Collectively, these avian constellations are known as "the southern birds."

Best visibility: September to November (14°N–90°S)

VOLANS

The Flying Fish

Volans, shortened form of the Latin name Piscis Volans ("the flying fish"), represents a so-called flying fish, of which there are more than sixty species. In order to escape predators, these fishes leap from the water and are able to glide above its surface long enough to give them the appearance of flying. The greatest concentration of them can be found in tropical and subtropical waters.

The faint constellation Volans is located in Southern Quadrant 2 and represents a flying fish coursing through the heavens with the constellation Dorado's mahi-mahi in pursuit. As is thematically appropriate for this watery scene, the constellation Carina (*the Keel*), which is part of the ancient constellation Argo Navis (*the Ship* Argo), is close by.

Best visibility: January to March (14°N–90°S)

CONSTELLATIONS AFTER PTOLEMY

HEVELIUS'S CONSTELLATIONS

CANES VENATICI

The Hunting Dogs

The constellation Canes Venatici, which means "the hunting dogs" in Latin, was imagined as being thematically linked to the ancient constellations Boötes (*the Herdsman*) and Ursa Major (*the Greater Bear*). The dogs are represented as a pair of greyhounds held on a leash by the driver of Boötes's oxcart, and they strain at the leash in pursuit of Ursa Major's bear.

Canes Venatici is located with Boötes in Northern Quadrant 3, while Ursa Major's bear lies in Northern Quadrant 2.

Best visibility: April to May (90°N–37°S)

LACERTA

The Lizard

As suggested by Hevelius himself, the faint constellation Lacerta (*the Lizard*) represents a specific species of lizard, the starred agama (*Laudakia stellio*), which, like the chameleon, has the ability to adapt the color of its skin to its environment as a camouflage. Lacerta's lizard creeps across the heavens in Northern Quadrant 4 between the ancient constellations Cygnus (*the Swan*) and Andromeda.

Best visibility: September to October (90°N–33°S)

LEO MINOR

The Lesser Lion

The indistinct constellation Leo Minor (*the Lesser Lion*) is smaller than the ancient constellation Leo (*the Lion*) and represents a lion cub. The lion cub strolls over the night sky in Northern Quadrant 2 between the ancient animal constellations Leo and Ursa Major (*the Greater Bear*).

Best visibility: March to April (90°N–48°S)

LYNX

The Lynx

The constellation Lynx is large but faint. Hevelius's original name for this constellation was an undecided one: Lynx, sive Tigris ("Lynx, or else Tiger"). He settled on Lynx. As Hevelius himself wrote, one would need the sharp eyes of a lynx to spot this faint constellation, a comment that apparently influenced his choice of name.

The lynx paces a large expanse of the night skies in Northern Quadrant 2 between the ancient constellations Auriga (*the Charioteer*) and Ursa Major (*the Greater Bear*). Nearby is the ancient constellation Gemini (*the Twins*), as well as several other animal constellations. The latter are the ancient constellations Cancer (*the Crab*) and Ursa Major, as well as the modern constellations Leo Minor (*the Lesser Lion*) and Camelopardalis (*the Giraffe*).

Best visibility: January to March (90°N–28°S)

SCUTUM

The Shield

The constellation Scutum (*the Shield*) was originally called Scutum Sobiescianum ("Sobieski's Shield") in honor of Hevelius's royal patron.

This small but distinct constellation lies in Southern Quadrant 4 and represents a shield hovering in the heavens between the ancient constellations Aquila (*the Eagle*) and Sagittarius (*the Archer*).

Best visibility: July to August (74°N–90°S)

SEXTANS

The Sextant

Sextans is a shortened form of the constellation's original name, Sextans Uraniae ("Sky Sextant"). This faint constellation lies on the Celestial Equator, its stars located in Northern Quadrant 2 and Southern Quadrant 2. The constellation Sextans represents the sextant that Hevelius himself used while making his naked-eye celestial observations and calculations. The sextant, which is roughly triangular in shape (but with a curved base), is located between the ancient constellations Leo (*the Lion*) and Hydra (*the Water Snake*).

Best visibility: March to April (78°N–83°S)

VULPECULA

The Fox

The constellation Vulpecula was originally called Vulpecula cum Ansere ("little fox with goose") in Latin, but a shorter version of the name was soon adopted. Although *vulpecula* literally means "little fox," the constellation is known simply as "the Fox" in English (*vulpes* in Latin).

Hevelius's fox trots across the heavens with an unlucky goose in its jaws. Vulpecula lies in Northern Quadrant 4 and is surrounded by the ancient constellations Pegasus, Cygnus (*the Swan*), and Delphinus (*the Dolphin*).

Best visibility: August to September (90°N–61°S)

CLASSICAL MYTHOLOGY OF THE CONSTELLATIONS

DE LACAILLE'S CONSTELLATIONS

ANTLIA

The Air Pump

Antlia is a shortened form of the name Antlia Pneumatica, "the air pump" in Latin. The longer Latin name is a translation of de Lacaille's original French name la Machine Pneumatique (literally "the pneumatic device"). De Lacaille's Antlia commemorates, and is depicted as, the air pump invented by French physicist and mathematician Denis Papin.

The constellation Antlia is located in Southern Quadrant 2 and is surrounded by the charismatic ancient constellations Centaurus (*the Centaur*), Hydra (*the Water Snake*), and Argo Navis (*the Ship* Argo).

Best visibility: March to April (49°N–90°S)

CAELUM

The Chisel

For de Lacaille, Caelum was originally les Burins, literally meaning "the chisels" in French. This French name became Caelum Sculptorium, Latin for "the sculptor or engraver's chisel," and was later shortened simply to Caelum, "the chisel."

De Lacaille's constellation Caelum commemorates and was depicted by him as a pair of instruments used by stonemasons, sculptors, and engravers. One was a burin, a tool with a mushroom-shaped handle attached to a long, pointed blade. The other was an echoppe, a needle-like instrument used in engraving and invented by the French printmaker Jacques Callot. The two instruments are shown crossing each other in an X form, securely tied.

The constellation Caelum is located in Southern Quadrant 1 and is surrounded by the ancient constellations Lepus (*the Hare*) and Eridanus (*the River*) as well as the modern constellations Columba (*the Dove*), Pictor (*the Painter's Easel*), Dorado (*the Golden Fish*), and Horologium (*the Pendulum Clock*).

Best visibility: December to January (41°N–90°S)

CARINA

The Keel

The stars constituting the constellation Carina (*the Keel*) were originally part of the ancient and enormous constellation Argo Navis (*the Ship* Argo). De Lacaille divided this large constellation into three parts: Vela (*the Sails*), Carina (*the Keel*), and Puppis (*the Stern*). Among the fourteen constellations attributed to him, the threefold division of Argo Navis counts as one.

Interestingly, de Lacaille's Carina represents the hull of the ship *Argo* and not its keel, which is in keeping with the original French name. He originally called his constellation Corps du Navire, literally "the ship's hull."

The constellation Carina is located in Southern Quadrant 2 and is logically adjacent to Vela and Puppis, the other parts of the ship.

Best visibility: January to April (14°N–90°S)

CIRCINUS

The Drawing Compass

Circinus (*the Drawing Compass*) is the smallest of the fourteen constellations introduced by de Lacaille, and he originally named it le Compas, "the compass" in French.

This constellation commemorates, and is depicted as, a pair of compasses. This drawing tool consists of two legs connected by a hinge allowing the distance between the legs to be adjusted. Drawing compasses are used to inscribe arcs or circles and to measure distances in technical applications such as mapping, navigation, drafting, and mathematics.

Circinus is located in Southern Quadrant 3 and is surrounded by the ancient constellations Centaurus (*the Centaur*) and Lupus (*the Wolf*) as well as the modern constellations Norma (*the Set Square*) and Triangulum Australe (*the Southern Triangle*).

Best visibility: May to June (19°N–90°S)

FORNAX

The Furnace

De Lacaille's constellation Fornax (*the Furnace*) represents a type of furnace used by chemists for distilling (separating) liquids. The furnace consists of a heating-box fitted with an alchemical still and an attached vessel for receiving condensed liquids. De Lacaille's original name for the constellation was le Fourneau Chymique (meaning "the chemical furnace"), which was translated into Latin as Fornax Chimiae and then shortened to Fornax.

The constellation Fornax is located in Southern Quadrant 1 and lies on the banks of the ancient river-constellation Eridanus (*the River*).

Best visibility: November to December (50°N–90°S)

HOROLOGIUM

The Pendulum Clock

For de Lacaille, Horologium (Latin for "a device to count the hours") was orignally l'Horloge à pendule & à secondes ("clock with pendulum and seconds hand") in French. The constellation was envisioned as representing a pendulum clock that, with the swing of its pendulum, could measure seconds as well as minutes and the hour. Lacaille used this very instrument to help him track the movements of stars.

The constellation Horologium is located in Southern Quadrant 1 and is surrounded by the ancient constellation Eridanus (*the River*) as well as the modern constellations Hydrus (*the Male Water Snake*), Reticulum (*the Reticle*), and Caelum (*the Chisel*).

Best visibility: November to December (23°N–90°S)

MENSA

The Table Mountain

Mensa is a shortened form of the name Mons Mensae ("the Table Mountain" in Latin) and was originally called Montagne de la Table (literally "Mountain of the Table") by de Lacaille.

De Lacaille's Mensa commemorates, and is depicted as, the particular mountain after which it is named, a distinctive flat-topped mountain overlooking Cape Town, South Africa, called Table Mountain. This mountain was, in fact, the location from which de Lacaille made his astronomical observations.

The constellation Mensa is located in Southern Quadrants 1 and 2 near the Southern Celestial Pole, and it is surrounded by the ancient constellation Eridanus (*the River*) as well as the modern constellations Hydrus (*the Male Water Snake*), Reticulum (*the Reticle*), Pictor (*the Painter's Easel*), and Caelum (*the Chisel*).

Best visibility: December to February (5°N–90°S)

MICROSCOPIUM

The Microscope

Microscopium, "the microscope" in Latin, is a classic example of a constellation named after a scientific instrument that was considered revolutionary in de Lacaille's time. His constellation Microscopium commemorates the compound microscope, an early optical microscope using visible light and multiple lenses to create magnified images of small objects. The microscope is represented as a cylinder, which contains the lenses, suspended over a boxy base.

This faint constellation is located in Southern Quadrant 4, tucked just under the forelegs of the ancient constellation Capricornus (*the Sea-Goat*). The ancient constellations Sagittarius (*the Archer*) and Piscis Austrinus (*the Southern Fish*) lie nearby, as do the modern constellations Telescopium (*the Telescope*), Indus (*the Indian*), and Grus (*the Crane*).

Best visibility: August to September (45°N–90°S)

NORMA

The Set Square

Another of de Lacaille's instrument constellations, Norma is a shortened form of the name Norma et Regula, "the set square and ruler" in Latin, which in turn is a translation of l'Équerre et la Règle, "the square and ruler" in French. De Lacaille viewed this constellation as an architect or surveyor's square and ruler, and for this reason, both implements appear in representations of the constellation.

The faint constellation Norma is located in Southern Quadrant 3 and is surrounded by the ancient constellations Ara (*the Altar*), Scorpius (*the Scorpion*), and Lupus (*the Wolf*) as well as the modern constellations Circinus (*the Drawing Compass*) and Triangulum Australe (*the Southern Triangle*).

Best visibility: June (29°N–90°S)

OCTANS

The Octant

De Lacaille's Octans commemorates a navigator's reflecting octant. This roughly triangular instrument was invented in 1730 by the English mathematician John Hadley (and, at the same time, by the lesser-known Philadelphia glazier Thomas Godfrey) to measure the height of a celestial body above the horizon. From this measurement, the latitudinal location of one's ship could be calculated. For de Lacaille, Octans (Latin) was originally l'Octans de Reflexion (literally "the reflective octant").

Located at the Southern Celestial Pole, the constellation Octans spreads through Southern Quadrants 1, 2, 3, and 4. It is surrounded by the constellations Chamaeleon (*the Chameleon*), Mensa (*the Table Mountain*), Hydrus (*the Male Water Snake*), Tucana (*the Toucan*), Indus (*the Indian*), Pavo (*the Peacock*), and Apus (*the Bird of Paradise*).

Best visibility: October (0°–90°S)

PICTOR

The Painter's Easel

Pictor, a shortened form of the name Pluteum Pictoris, "the painter's support," is one of several constellations de Lacaille named after tools or equipment used in the arts. His original French name for the constellation was le Chevalet et la Palette (literally "the easel and palette").

De Lacaille's Pictor represents an easel with a palette attached to it. The constellation is located in Southern Quadrants 1 and 2 and is surrounded by the ancient constellation Argo Navis (*the Ship* Argo) as well as the modern constellations Columba (*the Dove*), Caelum (*the Chisel*), Dorado (*the Golden Fish*), Reticulum (*the Reticle*), and Volans (*the Flying Fish*).

Best visibility: December to February (26°N–90°S)

PUPPIS

The Stern

Puppis, which means "the stern" in Latin, was originally named la Poupe du Navire (literally "ship's stern") by de Lacaille. The stars constituting Puppis were originally part of the ancient and enormous constellation Argo Navis (*the Ship* Argo). De Lacaille divided this large constellation into three parts: Vela (*the Sails*), Carina (*the Keel*), and Puppis (*the Stern*). Among the fourteen constellations attributed to him, the divided Argo Navis counts as one.

The constellation Puppis is located in Southern Quadrant 2 and is logically adjacent to Vela and Carina, the other parts of the ship.

Best visibility: January to February (39°N–90°S)

PYXIS

The Mariner's Compass

De Lacaille's Pyxis represents a magnetic compass and is one of the constellations that this astronomer introduced in homage to new and revolutionary scientific instruments. Originally named la Boussole ("the compass") in French, Pyxis is a shortened and slightly altered form of the name Pixis Nautica, "the nautical box" in Latin.

The small constellation Pyxis is located in Southern Quadrant 2 and lies between the ancient constellation Hydra (*the Water Snake*) and the modern constellation Puppis (*the Stern*), one of the three constellations into which the enormous ancient constellation Argo Navis (*the Ship* Argo) was divided.

Best visibility: February to March (52°N–90°S)

RETICULUM

The Reticle

Reticulum, literally "the little net" in Latin, is a translation of de Lacaille's original name for the constellation, le Reticule Rhomboide (literally "the rhomboid net"). The reticle was a small net in a telescope's eyepiece used to measure a star's position, and this constellation represents the reticle of de Lacaille's own telescope.

The small constellation Reticulum is located in Southern Quadrant 1 and is surrounded by the modern constellations Horologium (*the Pendulum Clock*), Hydrus (*the Male Water Snake*), Dorado (*the Golden Fish*), and Pictor (*the Painter's Easel*).

Best visibility: December (23°N–90°S)

CONSTELLATIONS AFTER PTOLEMY

SCULPTOR

The Sculptor

De Lacaille's constellation Sculptor depicts a sculptor's workshop complete with a sculptor's bench, tools, and a sculpted bust, which is in keeping with its original French name: l'Atelier du Sculpteur (literally "the sculptor's workshop"). The name Sculptor is an altered abbreviation of Apparatus Sculptoris, which is Latin for "the sculptor's tools."

The constellation Sculptor is located in Southern Quadrants 1 and 4, lying just south of the ancient constellations Aquarius (*the Water-Bearer*) and Cetus (*the Sea Monster*).

Best visibility: October to November (50°N–90°S)

TELESCOPIUM

The Telescope

For de Lacaille, Telescopium, the constellation's Latin name, was originally le Telescope, and depicts an aerial telescope. This type of telescope consisted of a light-gathering objective (in the form of lenses or mirrors) fixed to a long pole or tall structure. The objective was connected to the observer's eyepiece by a string or pole that allowed for the objective to be aimed toward various objects in the night sky.

The faint constellation Telescopium is located in Southern Quadrant 4 and is surrounded by the ancient constellations Sagittarius (*the Archer*), Corona Australis (*the Southern Crown*), Scorpius (*the Scorpion*), and Ara (*the Altar*) as well as the modern constellations Pavo (*the Peacock*), Indus (*the Indian*), and Microscopium (*the Microscope*).

Best visibility: July to August (33°N–90°S)

VELA

The Sails

The stars constituting Vela, "the sails" in Latin, were originally part of the ancient and enormous constellation Argo Navis (*the Ship* Argo). De Lacaille divided this large constellation into three parts: Vela, Carina (*the Keel*), and Puppis (*the Stern*). De Lacaille originally named his new constellation Voilure du Navire (literally "ship's sails"), and it is depicted as the sails of the ship *Argo*.

The constellation Vela is located in Southern Quadrant 2 and is logically adjacent to Carina and Puppis, the other parts of the ship.

Best visibility: February to April (32°N–90°S)

CONSTELLATIONS AFTER PTOLEMY

VOPEL'S CONSTELLATION

COMA BERENICES

The Hair of Berenice

The earliest depiction of the constellation Coma Berenices, Latin for "the hair of Berenice," was that of German cartographer and mathematician Caspar Vopel in 1536, who is credited with introducing this grouping of stars formally as a constellation. However, Coma Berenices's stars were a known grouping even in antiquity. Some ancient authors viewed its stars as the hair at the end of the lion Leo's tail, while others saw them as the hair of the god Dionysus's bride Ariadne resting under her crown, the ancient constellation Corona Borealis (*the Northern Crown*). Still others interpreted these stars as the Egyptian queen Berenice's hair. In this sense, the constellation is a very old one.

The following story, which is based on historical personages and events, was told about this starry head of hair in antiquity. Berenice II was the co-regent of Egypt (246–222 CE) and wife of her cousin Ptolemy III. When Ptolemy went to battle in the so-called Third Syrian War, Berenice made a sacred vow to the gods that she would offer them her hair as a gift of thanks if Ptolemy were to return safely. He did, and Berenice duly cut her locks, which were then presented to Aphrodite (or, by some accounts, all the gods). This event was commemorated in the literature of the day, and it was said that when Berenice's locks disappeared suddenly from Aphrodite's altar they reappeared as a grouping of stars in the sky.

Coma Berenice is a constellation of the Northern Celestial Hemisphere where it lies in Northern Quadrant 3. As indicated by how people in antiquity interpreted it, the constellation is located at the tail end of Leo (*the Lion*) and close to Corona Borealis, the starry representation of the princess Ariadne's crown.

Best visibility: April to May (90°N–56°S)

APPENDICES

CONSTELLATIONS BY THE SEASONS

A SUMMARY OF OPTIMAL VISIBILITY BY ASTRONOMICAL SEASON (WITH GLOBAL LATITUDES AT WHICH BEST TO VIEW THE CONSTELLATIONS)

NORTHERN WINTER, SOUTHERN SUMMER

DECEMBER TO JANUARY
Auriga (90°N–34°S)
Caelum (41°N–90°S)
Camelopardalis (90°N–3°S)
Dorado (20°N–90°S)
Eridanus (32°N–89°S)
Mensa (5°N–90°S)
Orion (79°N–67°S)
Pictor (32°N–90°S)
Reticulum (23°N–90°S)
Taurus (88°N–58°S)

JANUARY TO FEBRUARY
Auriga (90°N–34°S)
Camelopardalis (90°N–3°S)
Canis Major (56°N–90°S)
Canis Minor (89°N–77°S)
Carina (14°N–90°S)
Columba (46°N–90°S)
Gemini (90°N–55°S)
Lepus (62°N–90°S)
Lynx (90°N–28°S)
Mensa (5°N–90°S)
Monoceros (78°N–78°S)
Pictor (26°N–90°S)
Puppis (39°N–90°S)
Volans (14°N–90°S)

FEBRUARY TO MARCH
Camelopardalis (90°N–3°S)
Cancer (90°N–57°S)
Carina (14°N–90°S)
Chamaeleon (7°N–90°S)
Hydra (54°N–83°S)
Lynx (90°N–28°S)
Pyxis (52°N–90°S)
Ursa Major (90°N–16°S)
Vela (32°N–90°S)
Volans (14°N–90°S)

APPENDICES

NORTHERN SPRING, SOUTHERN AUTUMN

MARCH TO APRIL
Antlia (49°N–90°S)
Camelopardalis (90°N–3°S)
Carina (14°N–90°S)
Chamaeleon (7°N–90°S)
Crater (65°N–90°S)
Draco (90°N–4°S)
Hydra (54°N–83°S)
Leo (82°N–57°S)
Leo Minor (90°N–48°S)
Sextans (78°N–83°S)
Ursa Major (90°N–16°S)
Vela (32°N–90°S)

APRIL TO MAY
Camelopardalis (90°N–3°S)
Canes Venatici (90°N–37°S)
Centaurus (25°N–90°S)
Chamaeleon (7°N–90°S)
Coma Berenices (90°N–56°S)
Corvus (65°N–90°S)
Crux (25°N–90°S)
Draco (90°N–4°S)
Hydra (54°N–83°S)
Musca (14°N–90°S)
Ursa Major (90°N–16°S)
Virgo (67°N–75°S)

MAY TO JUNE
Apus (7°N–90°S)
Boötes (90°N–35°S)
Centaurus (25°N–90°S)
Circinus (19°N–90°S)
Draco (90°N–4°S)
Hydra (54°N–83°S)
Libra (60°N–90°S)
Lupus (34°N–30°S)
Ursa Minor (90°N–0°)
Virgo (67°N–75°S)

APPENDICES

NORTHERN SUMMER, SOUTHERN WINTER

JUNE TO JULY
Apus (7°N–90°S)
Ara (22°N–90°S)
Corona Borealis
  (90°N–50°S)
Draco (90°N–4°S)
Hercules (90°N–38°S)
Norma (29°N–90°S)
Ophiuchus
  (59°N–75°S)
Scorpius (44°N–90°S)
Serpens (74°N–64°S)
Triangulum Australe
  (19°N–90°S)

JULY TO AUGUST
Aquila (78°N–71°S)
Corona Australis
  (44°N–90°S)
Draco (90°N–4°S)
Hercules (90°N–38°S)
Lyra (90°N–42°S)
Pavo (15°N–90°S)
Sagittarius
  (44°N–90°S)
Scutum (74°N–90°S)
Serpens (74°N–64°S)
Telescopium
  (33°N–90°S)

AUGUST TO SEPTEMBER
Aquarius (65°N–86°S)
Capricornus
  (62°N–90°S)
Cygnus (90°N–28°S)
Delphinus
  (90°N–69°S)
Draco (90°N–4°S)
Indus (15°N–90°S)
Microscopium
  (45°N–90°S)
Pavo (15°N–90°S)
Sagitta (90°N–69°S)
Vulpecula (90°N–61°S)

NORTHERN AUTUMN, SOUTHERN SPRING

SEPTEMBER TO OCTOBER
Aquarius (65°N–86°S)
Cepheus (90°N–1°S)
Equuleus (90°N–77°S)
Grus (33°N–90°S)
Indus (15°N–90°S)
Lacerta (90°N–33°S)
Pegasus (90°N–53°S)
Piscis Austrinus
  (53°N–90°S)
Tucana (14°N–90°S)

OCTOBER TO NOVEMBER
Andromeda
  (90°N–37°S)
Cassiopeia (90°N–12°S)
Cetus (65°N–79°S)
Hydrus (8°N–90°S)
Octans (0°–90°S)
Phoenix (32°N–90°S)
Pisces (83°N–56°S)
Sculptor (50°N–90°S)
Tucana (14°N–90°S)

NOVEMBER TO DECEMBER
Aries (90°N–58°S)
Cassiopeia (90°N–12°S)
Cetus (65°N–79°S)
Eridanus (32°N–89°S)
Fornax (50°N–90°S)
Horologium
  (23°N–90°S)
Hydrus (8°N–90°S)
Perseus (90°N–31°S)
Triangulum
  (90°N–52°S)

APPENDICES

PTOLEMY'S 48 CONSTELLATIONS WITH THEIR ORIGINAL GREEK NAMES

(LITERALLY TRANSLATED)

ANDROMEDA: Andromeda (*Andromeda*)
AQUARIUS: Hydrokhoös (*Water Pourer*)
AQUILA: Aetos (*Eagle*)
ARA: Thymiatērion (*Incense Burner*)
ARGO NAVIS: Argō (*Argo*)
ARIES: Krios (*Ram*)
AURIGA: Hēniochos (*Charioteer*)
BOÖTES: Boōtēs (*Oxcart Driver*)
CANCER: Karkinos (*Crab*)
CANIS MAJOR: Kyōn (*Dog*)
CANIS MINOR: Prokyōn (*Preceding Dog*)
CAPRICORNUS: Aigokerōs (*Goat-Horned*)
CASSIOPEIA: Kassiepeia (*Cassiopeia*)
CENTAURUS: Kentauros (*Centaur*)
CEPHEUS: Kēpheus (*Cepheus*)
CETUS: Kētos (*Sea Monster*)
CORONA AUSTRALIS: Stephanos notios (*Southern Crown*)
CORONA BOREALIS: Stephanos (*Crown*)
CORVUS: Korax (*Crow*)
CRATER: Kratēr (*Mixing Bowl*)
CYGNUS: Ornis (*Swan*)
DELPHINUS: Delphin (*Dolphin*)
DRACO: Drakōn (*Dragon*)
EQUULEUS: Hippou protomē (*Horse's forequarters*)
ERIDANUS: Potamos (*River*)
GEMINI: Didymoi (*Twins*)
HERCULES: Engonasi (*Kneeler*)
HYDRA: Hydros (*Water Snake*)
LEO: Leōn (*Lion*)
LEPUS: Lagōos (*Rabbit*)
LIBRA: Khēlai (*Claws*)
LUPUS: Thērion (*Wild Beast*)
LYRA: Lyra (*Lyre*)
OPHIUCHUS: Ophioukhos (*Serpent Holder*)
ORION: Ōriōn (*Orion*)
PEGASUS: Hippos (*Horse*)
PERSEUS: Perseus (*Perseus*)
PISCES: Ikhthyes (*Fishes*)
PISCIS AUSTRINUS: Ikhthys notios (*Southern Fish*)
SAGITTA: Oistos (*Arrow*)
SAGITTARIUS: Toxotēs (*Archer*)
SCORPIUS: Skorpios (*Scorpion*)
SERPENS: Ophis (*Serpent*)
TAURUS: Tauros (*Bull*)
TRIANGULUM: Trigōnon (*Triangle*)
URSA MAJOR: Arktos megalē (*Great Bear*)
URSA MINOR: Arktos mikra (*Small Bear*)
VIRGO: Parthenos (*Maiden*)

APPENDICES

THE PRINCIPAL GODS OF THE GREEKS AND THEIR ROMAN EQUIVALENTS

Zeus (Jupiter)
Hera (Juno)
Poseidon (Neptune)
Demeter (Ceres)
Athena (Minerva)
Apollo (Apollo)
Artemis (Diana)

Ares (Mars)
Aphrodite (Venus)
Hephaestus (Vulcan)
Hermes (Mercury)
Hestia (Vesta)
Dionysus (Bacchus)

GLOSSARY OF ANCIENT SOURCES

NOTE: All ancient source materials have been translated by the author from the original languages.

ANTONINUS LIBERALIS (first or second century CE)—a grammarian and author of a collection of myths written in Greek and known as the *Tales of Metamorphosis*. The details of his life have not been preserved, and the *Tales of Metamorphosis* is his only surviving work.

APOLLODORUS (first or second century CE)—the name that, likely in error, has become associated with an encyclopedic summary in Greek of Greco-Roman myth and legend, entitled *Bibliotheca* (The Library).

APOLLONIUS OF RHODES (first half of the third century BCE)—author of the Greek epic poem *Argonautica* (Voyage of the *Argo*), centered on the hero Jason's quest for the golden fleece.

ARATUS (315?–240 BCE)—a Greek poet born in Cilicia, the southern coast of Asia Minor (modern Turkey). His only surviving work is a poem in 1,154 hexameter verses, entitled *Phaenomena* (Phenomena), on the positions, risings, settings, and mythology of the most important stars and constellations.

ARISTOTLE (384–322 BCE)—the philosopher Plato's most famous student and, later, a teacher of Alexander the Great, Aristotle founded the philosophical school in Athens known as the Academy. His extensive body of writings included works on ethics, logic, politics, poetry, biology, and natural science. Among his works on natural science are *Physica* (Physics), *De Caelo* (On the Heavens), and *Meteorologica* (Meteorology).

CATULLUS [GAIUS VALERIUS CATULLUS] (84?–55? BCE)—a Roman poet from the Italian town of Verona whose slim book of *Carmina* (Poems), a *libellus* ("little book"), as he called it, reveals that he was a member of the Roman Republic's "high society," and includes references to the orator and statesman Cicero as well as Julius Caesar and Caesar's rival the general Pompey the Great, among others.

DIODORUS (active circa 60–20 BCE)—known as Diodorus Siculus, "the Sicilian." He authored *The Library of History*, an extensive history of the known world from mythical times to Caesar's conquest of Gaul. His work, written in Greek, includes discussions of Egypt, Mesopotamia, India, Scythia, Arabia, North Africa, Greece, and Europe.

GLOSSARY OF ANCIENT SOURCES

HERODOTUS (480?–425 BCE)—known as the "father of history," he was the first person in the Western world to make the events of the past the subject of investigation. His history in Greek of the Greco-Persian Wars (490–479 BCE), *The Histories*, contains a wealth of geographical, mythological, political, and ethnographic information.

HESIOD (active circa 725 BCE)—according to Greek tradition, the author of two highly influential, instructional epic poems: the *Theogony*, which treats the origins of the universe and of the gods, and the *Works and Days*, which includes reflections on social and religious conduct as well as a farmer's calendar.

HOMER (eighth century BCE)—according to Greek tradition, author of the *Iliad* and *Odyssey*, together constituting the earliest extant examples of literature in the Western world. The so-called *Homeric Hymns*, a collection of poems celebrating the Greek gods and of unknown authorship, is wrongly attributed to him.

HYGINUS (second century CE?)—known, probably falsely, as the author of a handbook of mythology compiled from a variety of Greek sources and a manual of astronomy, also with mythological content: *Fabulae* (Stories) and *Astronomica* (Poetical Astronomy), respectively.

NONNUS (active third quarter of the fifth century CE?)—a Greek poet from the city of Panopolis (Akhmim) in Egypt. His epic poem *The Dionysiaca* (Things About Dionysus) centers on the life and exploits of the god Dionysus.

OVID [PUBLIUS OVIDIUS NASO] (43 BCE–18 CE)—among the best-known and highly acclaimed of Latin poets. His work includes the *Metamorphoses*, an epic poem that for centuries has been the primary source of Greek and Roman myth and legend. His other works include the controversial *Ars Amatoria* (The Art of Love), a manual on the arts of seduction; *Heroides* (The Heroines), a series of fictional letters in verse from heroines in mythology to their lovers; and *Fasti* (Calendar), a poetic work that goes through the official Roman calendar month by month, indicating festival days as well as their origins and mythology.

PAUSANIAS (circa 115–180 CE)—author of a description in Greek of mainland Greece based on his own travels. Cast as a travel guide, *The Description of Greece* provides a wealth of information regarding many now-lost sites, monuments, and artworks as well as the customs and beliefs of those regions that he visited.

PHILOSTRATUS (first to third centuries CE?)—presumed authors of a two-part work entitled *Eikones* (Images) in Greek and *Imagines* in Latin. Two members of the Philostratus family were possibly involved in authoring this work. *Eikones/Imagines* is a collection of detailed descriptions of works of art viewed by the author(s) in Naples.

GLOSSARY OF ANCIENT SOURCES

PLATO (428/7–348/7 BCE)—Athenian philosopher and founder of the philosophical community or school that came to be called the Academy. Among Plato's many writings is *The Republic*, a discourse on the ideal state, which features Socrates (by whom Plato had been deeply influenced) as a character.

PYTHAGORAS (circa 570–circa 495 BCE)—a Greek philosopher born on the island of Samos. He became famous for his contributions to mathematics and astronomy as well as his belief in transmigration of the soul.

SENECA [LUCIUS ANNAEUS SENECA] (4? BCE–65 CE)—a Roman statesman, philosopher, and dramatist. Born in Córdoba, Spain, Seneca was educated in Rome and became first tutor and then political adviser to the emperor Nero. Among his works are a group of tragedies on mythological themes: *Hercules furens* (The Madness of Hercules), *Troades* (The Trojan Women), *Phoenissae* (The Phoenician Women), *Medea*, *Phaedra*, *Oedipus*, *Agamemnon*, and *Thyestes*.

STRABO (circa 65 BCE–25 CE)—historian and geographer. He is known primarily for his wide-ranging work in Greek on geography, inclusive of Spain, Gaul, Italy, the Balkans, Asia Minor, India, Egypt, Northern Africa, and more. His seventeen-part work is known simply as *Geographia* (Geography).

THEOCRITUS (early third century BCE)—a Greek author of pastoral poems entitled the *Idylls*. Theocritus, who is called the creator of the bucolic genre of poetry, was likely born in Syracuse and then spent time working on the island of Cos as well as in Alexandria, Egypt. While thirty of the *Idylls* attributed to him have survived, not all were actually authored by him nor are all of the poems pastoral (dealing with the charms of country life) in content.

VERGIL [PUBLIUS VERGILIUS MARO] (70?–19 BCE)—illustrious author of the *Aeneid*, an epic poem recounting the founding of Rome and the origins of the Roman people. Vergil, who enjoyed the patronage of the emperor Augustus, was also the author of the *Eclogues*, a group of pastoral poems, and the *Georgics*, a didactic poem as much about agriculture as it is about the social and political concerns of the day.

MODERN SOURCES

Modern sources consulted include the following:

Dunkley, Jo. *Our Universe: An Astronomer's Guide.* Harvard University Press, 2020.
The International Astronomical Union website: https://www.iau.org.
Mitton, Jacqueline, ed. *Simply Astronomy.* DK Books, 2021.
Ridpath, Ian. *Stars and Planets.* DK Books, 2022.
Ridpath, Ian. *Star Tales.* James Clarke & Co. Ltd., 1989.
Toomer, Gerald J. *Ptolemy's "Almagest."* Princeton University Press, 1999.

ACKNOWLEDGMENTS

I owe an inestimable debt of gratitude to a number of people. I have to start by thanking Becky Koh, publishing director at Black Dog & Leventhal, for offering me the opportunity to write on two interrelated subjects that are dear to me: mythology and astronomy. This book, and its predecessor *Classical Mythology A to Z*, would never have seen the light of day were it not for Sarah Levitt, my amazing agent at Aevitas Creative. Jim Tierney has worked his magic in bringing the starry figures that populate the heavens so vividly to life. His are the figures that I will now always picture as I look up at the night sky.

As regards the book's production, I am enormously grateful to Melanie Gold and her team for their care and patience. For his generous feedback on portions of the text, I wish to thank astronomer Dr. Ian Griffin, director of Tūhura | Otago Museum, Dunedin, New Zealand.

Finally, I am grateful beyond words to my husband, Don Dunham, who read through several versions of the manuscript, endured countless debates on the book's optimal structure, and cheered me on throughout the writing process. And I am grateful to Muna and Kuri, *Canes Optimi*, for being a continual source of joy and inspiration.

INDEX

Main entry is in **bold**.
Illustrations are in *italics*.

Achelous, 84
Achilles
    Argo Navis and, 129
    Centaurus and, 149–150
    Iliad and, 4–5
    as inspiration for constellations, 3
    Sagittarius and, 187, 189
Acrisius, 35–36, 39
Admetus, 83, 114, 161, 163
Aeetes, 21, 23, 129, 131
Aeson, 129–130
Agamemnon, 24, 26, 97–98
Air Pump (Antila), **208**
Alcestis, 83
Alcmene, 77–79
Alexander the Great, 6
Alexandria, 6
Almagest, xiii, 2, 5–9, 72, 160, 194
Altar (Ara), **146–148**, *146*, 160
Amalthea, 24, 28
Ammon, 29, 95
Ammon, oracle of, 19, 31, 95
Amphitrite, 100, 102
Amphitryon, 54, 56, 77–79, 136
Amulius, 158, 160–161
Amycus, 57, 59
Anaximenes, 7
Andromeda
    Cassiopeia and, 29, 31
    Cepheus and, 95, 97
    Cetus and, 119–120
    constellation of, 10, **18–20**
    illustration of, *18*
    Pegasus and, 109
    Perseus and, 35, 39–40
    surrounding constellations and, 12
Andromeda Galaxy, 20
Antila (Air Pump), **208**
Antoninus Pius, 6
Aphrodite
    Capricornus and, 179, 181
    Coma Berenices and, 217
    Corona Borealis and, 71–72
    Pisces and, 40, 42
    Piscis Austrinus and, 184, 186
Apollo
    Aquila and, 93–94
    Corvus and, 153, 155
    Crater and, 137, 139
    Hercules and, 77
    Lyra and, 105–106, 108
    Ophiuchus and, 161, 163
    Perseus and, 35, 39
    Roman name for, 223
    Sagitta and, 112, 114
Apollodorus, 5
Apus (Bird of Paradise), **198**
Aquarius (Water-Bearer), 11, 94, **176–179**, *176*
Aquila (Eagle), **92–94**, *92*, 115, 148, 177
Ara (Altar), 93, **146–148**, *146*, 160
Arcas, 63–64, 69, 70–71, 86, 88, 158, 160
Archer (Sagittarius), 152, 182, **187–190**, *188*
Arctophylax, 70–71
Ares, 21, 23, 42, 83, 223
Argo, 21, 23, 59, 82, 106, 130–132

INDEX

Argo Navis (Ship *Argo*), 106, **128**–**132**, *128*, 195, 209, 213, 216
Argonautica (Apollonius of Rhodes), 74
Argonauts, 21, 59, 106, 129–132, 187, 189
Argus, 129–130
Ariadne, 43, 71–72, 74, 217
Aries (Ram)
    Centaurus and, 150
    constellation of, **21–23**
    Gemini and, 59
    Hercules and, 82
    illustration of, 22
    Jason and, 130
    Triangulum and, 48
    Zodiac constellations and, 11
Arion, 100, 102–103
Aristaeus, 105–106
Aristotle, 4, 7
Arrow (Sagitta), **112**–**115**, *113*, 142, 163
Artemis
    Callisto and, 13, 98
    Canis Major and, 133–134
    Canis Minor and, 54, 56
    Gemini and, 57, 59
    Hercules and, 77, 81–82
    Lepus and, 123
    Libra and, 155–156
    Orion and, 31, 34, 123
    Roman name for, 223
    Scorpius and, 164, 166
    Ursa Major and, 63
    Ursa Minor and, 86
Asclepius, 112, 114, 153, 161, 163–164, 166, 168
Astraea, 155–156, 169, 171, 172
Astraeus, 169, 171
Atalanta, 57, 59
Athamas, 21, 23, 24
Athena
    Andromeda and, 19–20
    Aquarius and, 177–178
    Argo Navis and, 129–130, 132

Auriga and, 28
Cassiopeia and, 29, 31
Cepheus and, 95, 97
Cetus and, 119–120
Equuleus and, 103, 105
Pegasus and, 109, 111
Perseus and, 35, 38, 40
Roman name for, 223
Taurus and, 45
Atlas, 5, 74, 76, 84
Augeas, 77, 82
Augustus, 6
Aura, 133, 136
Auriga (Charioteer), **24**–**28**, *27*

Babylonian star catalogues, 12
Bacchantes, 105, 108
Bebryces, 59, 131
Bellerophon, 103, 109, 111–112, 140, 142
Bird of Paradise (Apus), **198**
Boötes (Herdsman), 56, **68**–**71**, *68*, 134, 171, 205
Brahe, Tycho, 8
brightness, indications of, 10
Bull (Taurus), 5, 11, **43**–**46**, *44*, 83, 98, 136, 168

Caelum (Chisel), **208–209**
Callisto
    Boötes and, 69, 70
    Cygnus and, 97–98
    as inspiration for constellations, 3
    Lupus and, 158, 160
    punishment of, 13
    Ursa Major and, 63–64
    Ursa Minor and, 86, 88
Callot, Jacques, 208
Camelopardalis (Giraffe), **196**, 197
Cancer (Crab), 11, 12, **52**–**53**, *52*, 81, 140, 142
Canes Venatici (Hunting Dogs), 71, **205**
Canicula (Dog Star), 70, 136, 171

INDEX

Canis Major (Greater Dog)
    Canis Minor and, 54, 56
    constellation of, **133–137**
    illustration of, *135*
    as inspiration for constellations, 12
    Lepus and, 123
    Maera and, 70
    Orion and, 34
    Virgo and, 171
Canis Minor (Lesser Dog), 34, **54–56**, *55*, 70, 123, 136, 171
Capricornus (Sea-Goat), 11, 12, **179–181**, *180*
Carina (Keel), 132, 195, **209**, 213, 216
Cassiopeia
    Andromeda and, 19
    Cepheus and, 95, 97
    Cetus and, 119
    constellation of, **29–31**
    illustration of, *30*
    as inspiration for constellations, 12
    Perseus and, 35, 40
Castor, 57, 59, 77, 79, 98, 103, 105, 129–130, 149–150
Cecrops, 24, 28, 177–179
Cedalion, 31, 34
Celestial Equator, 10
Celestial Quadrants, 9, 10
Celestial Sphere, 9, 10–11
Centaur (Centaurus), 130, 139, 142, 148, **149–152**, *151*, 153, 163, 187, 197
Centaurus (Centaur), 12, 82, 85, 130, 139, 142, 148, **149–152**, *151*, 153, 163, 187, 197
Cephalus, 54, 56, 133, 136
Cepheus
    Andromeda and, 19–20
    Cassiopeia and, 29, 31
    Cetus and, 119
    constellation of, **95–97**
    illustration of, *96*
    as inspiration for constellations, 12
    Perseus and, 35, 39, 40

Cerberus, 60, 62, 74, 76, 77, 84, 105, 108, 140, 142
Ceryneian Hind, 77, 81–82
Cetus (Sea Monster)
    Andromeda and, 19–20
    Cassiopeia and, 29, 31
    Cepheus and, 95, 97
    constellation of, **118–120**
    illustration of, *118*
    as inspiration for constellations, 12
    Perseus and, 35, 39, 40
Chamaeleon (Chameleon), **199**, 201
Chameleon (Chamaeleon), **199**, 201
Charioteer (Auriga), **24–28**, *27*
Charon, 105–106, 108
Charybdis, 129, 132
Chimera, 62, 76, 103, 111, 140, 142
Chiron
    Ara and, 147
    Argo Navis and, 129–130
    Centaurus and, 149–150, 152
    Corvus and, 153
    Hercules and, 77, 79, 82, 142
    Ophiuchus and, 161, 163
    Sagittarius and, 187, 189
Chisel (Caelum), **208–209**
Chrysaor, 35, 38, 77, 83, 109, 111
Circe, 129, 132
Circinus (Drawing Compass), **209**
Cleopatra, 6
Clymene, 24, 26, 120, 122
Clytemnestra, 98
Columba (Dove), **196**
Coma Berenices (Hair of Berenice), 195, 217
Copernicus, Nicolaus, 8
Corona Australis (Southern Crown), 72, **182–184**, *183*
Corona Borealis (Northern Crown), **71–74**, *73*, 102, 217
Coronis, 153, 161, 163
Corvus (Crow), 139, **152–155**, *154*, 163

INDEX

Crab (Cancer), 11, 12, **52–53**, *52*, 81, 140, 142
Crane (Grus), **200**
Crater (Cup), **137–140**, *138*, 153, 155
Creon, 54, 56
Cretan Bull, 43, 45, 77, 83
Cronus
    Aquila and, 93
    Ara and, 147–148
    Centaurus and, 149–150
    Sagitta and, 112, 114
    Sagittarius and, 187, 189
    Triangulum and, 46, 48
    Ursa Major and, 64
    Ursa Minor and, 86, 88
Crotus, 187, 189–190
Crow (Corvus), 139, **152–155**, *154*, 163
Crux (Southern Cross), **197**
Cup (Crater), **137–140**, *138*, 153, 155
Cupid, 40, 42, 179, 181
Cyclopes, 93, 112, 114, 147–148, 161, 163
Cygnus (Swan), 57, **97–99**, *99*, 115

Daedalus, 43, 45
Danaë, 35–36
Dardanelles, 23
de Houtman, Frederick, 194, 203
de Lacaille, Nicolas-Louis, 132, 194, 195, 202, 208–216
Deianeira, 77, 84–85, 149–150, 152
Delphi, oracle at, 21, 23, 36, 79, 114
Delphinus (Dolphin), **100–103**, *101*
Demeter
    Auriga and, 24–25
    Boötes and, 69, 70
    Roman name for, 223
    Triangulum and, 46, 48
    Ursa Minor and, 86, 88
    Virgo and, 169, 172
Demophon, 137, 139–140
Deucalion, 177–179
Dictys, 35–36, 39

Dikē, 155–156, 169
Dionysus
    Boötes and, 69–70
    Canis Major and, 133–134
    Canis Minor and, 54, 56
    Centaurus and, 149, 152
    Coma Berenices and, 217
    Corona Australis and, 182, 184
    Corona Borealis and, 71–72
    Delphinus and, 100, 102
    Lyra and, 105, 108
    Roman name for, 223
    Virgo and, 169, 171
Dog Star (Canicula), 70, 136, 171
Dolphin (Delphinus), **100–103**, *101*
Dorado (Golden Fish), **199**
Dove (Columba), **196**
Draco (Dragon)
    constellation of, **74–76**
    Hercules and, 84, 85
    Hydra and, 142
    illustration of, *75*
    Leo and, 62
    Pegasus and, 111
    Pisces and, 42
    Sagitta and, 115
Dragon (Draco)
    constellation of, **74–76**
    Hercules and, 84, 85
    Hydra and, 142
    illustration of, *75*
    Leo and, 62
    Pegasus and, 111
    Pisces and, 42
    Sagitta and, 115
Drawing Compass (Circinus), **209**

Ea, 12
Eagle (Aquila), **92–94**, *92*, 115, 148, 177
Echidna, 60, 62, 74, 76, 140, 142
Eos, 133, 136, 169, 171
Equuleus (Foal), **103–105**, *104*

INDEX

Erichthonius, 24–25, 28
Eridanus (River), 120–122, 121
Erigone, 56, 69–70, 133–134, 169, 171
Erinyes, 35–36
Eros, 40, 42
Erymantian Boar, 77, 82
Eudoxus, 7
Eupheme, 187, 189
Europa, 43, 45, 97–98, 133–134, 136
Euryale, 31–32
Eurydice, 105–106, 108
Eurystheus, 60, 62, 77, 79, 81, 82, 83, 84

Fabulae (Hyginus), 140
Firmamentum Sobiescianum sive Uranographia ("Sobieski's Heaven, or Star Catalogue"), 194
Fishes (Pisces), 11, 40–43, 41, 181, 186
Five Good Emperors, 6
Fly (Musca), 199, 201
Flying Fish (Volans), 199, 204
Foal (Equuleus), 103–105, 104
Fornax (Furnace), 210
Fortuna, 169, 171
Fox (Vulpecula), 207
Furnace (Fornax), 210

Gaia
    Ara and, 147
    celestial phenomena and, 3
    Draco and, 74
    Hercules and, 84
    Hesiod on, 5
    Lepus and, 123
    Libra and, 155–156
    Orion and, 32, 34, 134
    Pisces and, 40, 42
    Sagitta and, 112, 114
    Scorpius and, 164
Galactic Quadrants, 9
Ganymede, 93–94, 177–178
Garden of the Hesperides, 62, 74, 84, 115

Gemini (Twins)
    Argo Navis and, 130
    Cancer and, 53
    Centaurus and, 150
    constellation of, 57–60
    Cygnus and, 98
    Equuleus and, 105
    Hercules and, 77, 79
    illustration of, 58
    Zodiac constellations and, 11
Georgics (Vergil), 103
Gerard of Cremona, 8
Geryon, 83
Giraffe (Camelopardalis), 196, 197
Glaucus, 166, 168
Godfrey, Thomas, 212
Golden Fish (Dorado), 199
Golden Fleece, 131
Gorgons, 38, 109, 111
Graiae, 35, 38
Greater Bear (Ursa Major), 5, 63–65, 65, 70, 98, 160, 205
Greater Dog (Canis Major)
    Canis Minor and, 54, 56
    constellation of, 133–137
    illustration of, 135
    as inspiration for constellations, 12
    Lepus and, 123
    Maera and, 70
    Orion and, 34
    Virgo and, 171
Greek constellation names, original, 222–223
Grus (Crane), 200

Hades
    Hercules and, 77, 83, 84
    Lyra and, 106, 108
    Persephone and, 48, 172
    Triangulum and, 46
    Ursa Minor and, 86, 88
Hadley, John, 212

INDEX

Hair of Berenice (Coma Berenices), 217
Hare (Lepus), 12, 34, **123–125**, *125*
Harpies, 129, 131
Hebe, 85
Hecatoncheires, 112, 114, 147
Helen of Troy, 57, 59, 98
Heliades, 120, 122
Helios
    Aquila and, 93
    Argo Navis and, 129
    Auriga and, 24, 26, 28
    celestial phenomena and, 3
    Eridanus and, 120, 122
    Medea and, 132
    Orion and, 34
Helle, 21, 23
Hellespont, 23
Hephaestus, 24, 28, 223
Hera
    Argo Navis and, 129–130
    Boötes and, 69, 70
    Callisto and, 3
    Cancer and, 53
    Canis Major and, 133, 136
    Canis Minor and, 54
    Centaurus and, 150
    Corona Australis and, 182
    Cygnus and, 98
    Draco and, 74, 76
    Hercules and, 77–79, 83, 84, 85
    Hydra and, 140, 142
    as inspiration for constellations, 12–13
    Leo and, 60, 62
    Orion and, 32
    Roman name for, 223
    Ursa Major and, 63–64
    Ursa Minor and, 86, 88
Heracles, 11
Hercules
    Aquila and, 93–94
    Ara and, 147–148
    Argo Navis and, 129–130
    Auriga and, 26
    Cancer and, 53
    Centaurus and, 149–150, 152
    constellation of, **77–85**
    Crater and, 137, 139
    Draco and, 74, 76
    Gemini and, 57, 59
    Greek name for, 11
    Hydra and, 140, 200
    illustration of, *80*
    as inspiration for constellations, 12–13
    Leo and, 60, 62
    Lyra and, 106, 108
    Pisces and, 40, 42
    Sagitta and, 114–115
    Sagittarius and, 187, 189
    Taurus and, 43, 45
Herdsman (Boötes), 56, **68–71**, *68*, 134, 171, 205
Hermes
    Aries and, 21, 23
    Auriga and, 24–25
    Capricornus and, 179, 181
    Libra and, 155–156
    Lupus and, 158, 160
    Lyra and, 106
    Orion and, 32
    Perseus and, 35, 38
    Roman name for, 223
    Triangulum and, 46, 48
    Ursa Minor and, 86
Herodotus, 5
Hesiod, 4, 5
Hesperides, 74, 76, 77, 84, 112, 115, 142
Hesperus, 77, 84
Hestia, 86, 88, 223
Hevelius, Johannes, 194, 205–207
Hipparchus, 7
Hippodamia, 24–25, 35–36
Hippolyta, 24, 26, 77, 83
Hippolytus, 24–25, 26
Homer, 4, 5

INDEX

Horologium (Pendulum Clock), 210
Horus, 184, 186
Hunting Dogs (Canes Venatici), 71, 205
Hyades, 5
Hydra (monster)
    Cancer and, 53
    Centaurus and, 152
    Corvus and, 153
    Draco and, 76
    Hercules and, 77, 81
    Leo and, 60, 62
    Pegasus and, 111
    Pisces and, 40, 42
    Sagitta and, 112, 114
    Sagittarius and, 187, 189
Hydra (Water Snake), 12, **140–142**, *141*, 153, 155
Hydrus (Male Water Snake), **200**
Hyginus, works of, 5
Hylas, 78, 82
Hyrieus, 32

Iasion, 69, 70
Icarius, 54, 56, 69–70, 133–134, 169, 171
Iliad (Homer), 4
Indian (Indus), **201**
Indus (Indian), **201**
Ino, 21, 23
International Astronomical Union (IAU), 9–10, 132, 194, 195
Io, 98
Iobates, 109, 111
Iolaus, 78, 142
Iphicles, 57, 78, 79
Ischys, 153
Isis, 184, 186
Ixion, 149–150

Jason
    Argo Navis and, 129–132
    Aries and, 21, 23
    Centaurus and, 149–150

Gemini and, 57, 59
Hercules and, 82
Lyra and, 106
Sagittarius and, 187, 189
Jupiter, 7

Keel (Carina), 132, 195, **209**, 213, 216
Kepler, Johannes, 8–9
Keyser, Pieter Dirkszoon, 194, 203

Lacerta (Lizard), **205**
Ladon the dragon
    Draco and, 74, 76
    Hercules and, 12, 78, 84, 85
    Leo and, 62
    Pegasus and, 111
    Pisces and, 40, 42
    Sagitta and, 112, 115
Laelaps, 54, 56, 133–134, 136
Laocoön, 161, 163
Leda, 57, 98
Leo (Lion)
    Cancer and, 53
    Coma Berenices and, 217
    constellation of, **60–62**
    Draco and, 76
    Hercules and, 12, 81
    illustration of, *61*
    Leo Minor and, 206
    Pegasus and, 111
    Pisces and, 42
    Zodiac constellations and, 11
    See also Nemean Lion
Leo Minor (Lesser Lion), **206**
Lepus (Hare), 12, 34, **123–125**, *125*
Lesser Bear (Ursa Minor), 64, 71, **86–88**, *87*, 160
Lesser Dog (Canis Minor), 34, **54–56**, *55*, 70, 123, 136, 171
Lesser Lion (Leo Minor), **206**
Libra (Scales), 11, 34, **155–158**, *157*, 166, 172
Linus, 79, 108

INDEX

Lion (Leo)
    Cancer and, 53
    Coma Berenices and, 217
    constellation of, 60–62
    Draco and, 76
    Hercules and, 12, 81
    illustration of, *61*
    Leo Minor and, 206
    Pegasus and, 111
    Pisces and, 42
    Zodiac constellations and, 11
    See also Nemean Lion
Little Dipper, 88
Lizard (Lacerta), **205**
Luna, 3
Lupus (Wolf), 88, 148, **158–161**, *159*
Lycaon, 63, 86, 88, 158, 160
Lynx, **206**
Lyra (Lyre), **105–108**, *107*, 108, 130
Lyre (Lyra), **105–108**, *107*, 108, 130

Maera, 54, 56, 69–70, 133–134, 171
Male Water Snake (Hydrus), **200**
Mares of Diomedes, 78, 83
Mariner's Compass (Pyxis), **214**
Mark Antony, 6
Mars, 7, 40, 158, 160
Matusius, 137, 140
Medea, 129, 131–132
Medusa
    Andromeda and, 19–20
    Canis Major and, 133
    Cassiopeia and, 29
    Cepheus and, 95, 97
    Cetus and, 119–120
    Equuleus and, 103
    Hercules and, 78
    Pegasus and, 109, 111
    Perseus and, 35, 38, 40
Megara, 78, 79
Meleager, 57, 59
Menelaus, 98

Mensa (Table Mountain), 194–195, **211**
Mercury, 7
Metamorphoses (Ovid)
    gods' behavior in, 13
    as source for myths, 5
Microscope (Microscopium), **211**
Microscopium (Microscope), 194, **211**
Milky Way, 26, 79, 98
Minos, 43, 45, 71–72, 82–83, 133, 136, 166, 168
Minotaur, 43, 45, 71–72, 74, 82–83, 166, 168
Monoceros (Unicorn), **197**
Moon, 5, 7
Musca (Fly), 199, **201**
Muses, 103, 106, 108, 111, 187, 189–190
Museum (Place of the Muses), 6
Myrtilus, 24–26

Nemean Lion
    Cancer and, 53
    Draco and, 74, 76
    Hercules and, 12, 78, 79, 81
    Hydra and, 140, 142
    Leo and, 60, 62
    Pegasus and, 109, 111
    Pisces and, 42
Nephele, 21, 23, 149–150
Nereids, 19, 29, 95
Nessus, 78, 84–85, 142, 149–150, 152
Noah, 196
Nonnus, 5
Norma (Set Square), **212**
North Star (Polaris), 88
Northern Crown (Corona Borealis), **71–74**, *73*, 102, 217
northern hemisphere constellation chart, x
Numitor, 158, 160–161

Ocean, 5
Oceanus, 24, 26, 150
Octans (Octant), **212**

237

INDEX

Octant (Octans), **212**
Odysseus, 114
Oenomaus, 24–26
Oenopion, 32, 34
On the Revolutions of the Celestial Spheres (Copernicus), 8
Ophiuchus (Serpent-Holder), 114, 153, **161–164**, *162*
Orestes, 24, 26
Orion
    Achilles's shield and, 5
    Canis Major and, 133–134
    Canis Minor and, 54, 56
    constellation of, **31–34**
    illustration of, *33*
    as inspiration for constellations, 3, 12
    Lepus and, 123
    Libra and, 155–156, 158
    Scorpius and, 164, 166
Orpheus, 79, 106, 108, 129–130
Osiris, 186
Ouranos, 5
Ovid
    gods' behavior and, 13
    works of, 5

Painter's Easel (Pictor), **213**
Pan, 179, 181, 187, 189
Panathenaia, 28
Papin, Denis, 208
Paris, 98
Pasiphaë, 43, 45, 71–72
Pausanias, 5
Pavo (Peacock), **202**
Peacock (Pavo), **202**
Pegasus
    constellation of, **109–112**
    Equuleus and, 103
    Hercules and, 78, 83
    Hydra and, 140, 142
    illustration of, *110*
    Perseus and, 35, 38, 40

Pelias, 129–132
Pelops, 24–26
Pendulum Clock (Horologium), **210**
Periphas, 93–94
Persephone, 24–25, 46, 48, 78, 84, 106, 108, 172
Perses, 35, 40
Perseus
    Andromeda and, 19–20
    Canis Major and, 133, 136
    Cassiopeia and, 29, 31
    Cepheus and, 97
    Cetus and, 119–120
    constellation of, **35–40**
    Hercules and, 78
    illustration of, *37*
    as inspiration for constellations, 12
    Pegasus and, 109, 111
Phaedra, 24, 26
Phaedra (Seneca), 97
Phaenomena (Aratus), 164
Phaethon, 24–25, 26, 28, 120, 122
Philoctetes, 78, 85
Philomelus, 69, 70
Philyra, 149–150, 187, 189
Phineus, 20, 129, 131
Phoenix, **202**
Pholus, 82, 137, 139, 142, 148, 149, 152, 187, 189
Phrixus, 21, 23
Pictor (Painter's Easel), **213**
Pisces (Fishes), 11, **40–43**, *41*, 181, 186
Piscis Austrinus (Southern Fish), **184–186**, *185*, 200
Plancius, Petrus, 194, 196–204
Plato, 4
Pleiades, 5
Plutus, 69, 70
Polaris (North Star), 88
Pollux, 57, 59, 98, 103, 105, 129–131, 149–150
Polydectes, 35–36, 39

238

INDEX

Poseidon
    Andromeda and, 19
    Aquarius and, 177–178
    Auriga and, 24–25, 26
    Cassiopeia and, 29, 31
    Cepheus and, 95
    Cetus and, 119–120
    Corona Borealis and, 71–72
    Delphinus and, 100, 102
    Equuleus and, 103, 105
    Libra and, 155–156
    Orion and, 32
    Pegasus and, 109, 111
    Perseus and, 38
    Phineus and, 131
    Roman name for, 223
    Taurus and, 43, 45
    Triangulum and, 46, 48
    Ursa Minor and, 86, 88
Procris, 133, 136
Prometheus, 93–94, 114, 115, 177
Ptolemy, xiii, 2, 6, 7–9, 72, 156, 160, 194, 195
Ptolemy III, 217
Puppis (Stern), 132, 195, 209, **213**, 216
Pyrrha, 177–178
Pythagoras, 4, 7–8
Python, 114
Pyxis (Mariner's Compass), **214**

Quadrants, 9

Ram (Aries)
    Centaurus and, 150
    constellation of, **21–23**, *23*
    Gemini and, 59
    Hercules and, 82
    illustration of, *22*
    Jason and, 130
    Triangulum and, 48
    Zodiac constellations and, 11
Remus, 158, 160–161

Reticle (Reticulum), **214**
Reticulum (Reticle), **214**
Rhea, 64, 86, 88, 148, 150, 187, 189
Rhea Silvia, 158, 160
River (Eridanus), **120–122**, *121*
Roman names for gods, 223
Romulus, 158, 160–161

Sagitta (Arrow), 12, 81, 94, **112–115**, *113*, 142, 163
Sagittarius (Archer), 11, 152, 182, **187–190**, *188*
Sails (Vela), 132, 195, 209, 213, **216**
Saturn, 7
Scales (Libra), **155–158**, *157*, 166, 172
Scorpion (Scorpius)
    Canis Minor and, 56
    constellation of, **164–166**
    illustration of, *165*
    Lepus and, 123
    Libra and, 155–156, 158
    Orion and, 34, 134
    Zodiac constellations and, 11
Scorpius (Scorpion)
    Canis Minor and, 56
    constellation of, **164–166**
    illustration of, *165*
    Lepus and, 123
    Libra and, 155–156, 158
    Orion and, 34, 134
    Zodiac constellations and, 11
Sculptor, **215**
Scutum (Shield), **206**
Scylla, 129, 132
Sea Monster (Cetus), **118–120**, *118*
Sea-Goat (Capricornus), 11, 12, **179–181**, *180*
seasonal shifts, 10, 220–222
Selene, 3
Semele, 100, 102, 182, 184
Serpens (Snake), 163, **166–168**, *167*
Serpens Caput (Serpent's Head), 168

239

INDEX

Serpens Cauda (Serpent's Tail), 168
Serpent-Holder (Ophiuchus), 114, 153, 161–164, *162*
Serpent's Head (Serpens Caput), 168
Serpent's Tail (Serpens Cauda), 168
Set Square (Norma), **212**
Seth, 186
Sextans (Sextant), **207**
Sextant (Sextans), **207**
Shield (Scutum), **206**
Ship *Argo* (Argo Navis), 106, **128**–**132**, *128*, 209, 213, 216
Side, 32
Sirius (Dog Star), 70, 136
Snake (Serpens), 163, **166**–**168**, *167*
Sobieski, Jan III, 194
sources
    ancient, 224–226
    modern, 227
Southern Cross (Crux), **197**
Southern Crown (Corona Australis), **182**–**184**, *183*
Southern Fish (Piscis Austrinus), **184**–**186**, *185*, 200
southern hemisphere constellation chart, xi
Southern Triangle (Triangulum Australe), **203**
St. Elmo's fire, 60
Stephanos, 72
Stern (Puppis), 132, 195, 209, **213**, 216
Stheneboea, 109, 111
Strabo, 5
Stymphalian Birds, 78, 82, 115
Sun, 5, 7, 11
Swan (Cygnus), 57, **97**–**99**, *99*, 115
Syrinx, 179, 181

Table Mountain (Mensa), **211**
Talus, 130, 132
Tantalus, 24–25
Taurus (Bull), 5, 11, **43**–**46**, *44*, 83, 98, 136, 168

Telescope (Telescopium), 202, **215**
Telescopium, 194
Telescopium (Telescope), 202, **215**
Teumessian Fox, 54, 56
Themis, 169, 171, 177–178
Theogony (Hesiod), 5
Theseus, 25, 26, 43, 45, 59, 71–72
Third Syrian War, 217
Thyestes, 25, 26
Titans, Zeus and, 48, 93, 147–148, 181. *See also* Astraeus; Atlas; Cronus; Oceanus; Prometheus; Rhea
Toucan (Tucana), **203**
Triangle (Triangulum), **46**–**49**, *47*
Triangulum (Triangle), **46**–**49**, *47*
Triangulum Australe (Southern Triangle), 46, **203**
Trojan Horse, 163
Trojan War, 4, 26, 98, 163
Tucana (Toucan), **203**
Twelve Labors of Hercules, 62, 76, 79, 114–115. *See also* Hercules; *individual labors*
Twins (Gemini)
    Argo Navis and, 130
    Cancer and, 53
    Centaurus and, 150
    constellation of, **57**–**60**
    Cygnus and, 98
    Equuleus and, 105
    Hercules and, 77, 79
    illustration of, *58*
    Zodiac constellations and, 11
Tyche, 169, 171, 172
Tyndareus, 57, 98
Typhon
    Capricornus and, 179, 181
    Draco and, 74, 76
    Hydra and, 140, 142
    Leo and, 60, 62
    Pisces and, 42

Unicorn (Monoceros), **197**

INDEX

Uranus, 112, 114, 147
Ursa Major (Greater Bear), 5, **63–65**, *65*, 70, 98, 160, 205
Ursa Minor (Lesser Bear), 64, 71, **86–88**, *87*, 160

Vela (Sails), 132, 195, 209, 213, **216**
Venus, 7
Vesta, 158, 160
Vestal Virgins, 160
Virgin (Virgo)
    Boötes and, 70
    Canis Major and, 134
    Canis Minor and, 56
    constellation of, **169–172**
    illustration of, *170*
    Libra and, 156, 158
    Zodiac constellations and, 11
Virgo (Virgin)
    Boötes and, 70
    Canis Major and, 134
    Canis Minor and, 56
    constellation of, **169–172**
    illustration of, *170*
    Libra and, 156, 158
    Zodiac constellations and, 11
Volans (Flying Fish), 199, **204**
Vopel, Caspar, 194, 195, 217
Vulpecula (Fox), **207**

Water Snake (Hydra), **140–142**, *141*, 153, 155
Water-Bearer (Aquarius), **176–179**, *176*
Wolf (Lupus), 88, 148, **158–161**, *159*
writing, history of, 4

Zeus
    Andromeda and, 19
    Aquarius and, 177–178
    Aquila and, 93–94
    Ara and, 147–148
    Argo Navis and, 130
    Aries and, 21, 23
    Auriga and, 26
    Boötes and, 69–70
    Canis Major and, 133–134, 136
    Canis Minor and, 54, 56
    Capricornus and, 179, 181
    Centaurus and, 149
    Corona Australis and, 182, 184
    Corvus and, 153
    Crater and, 137, 139
    Cygnus and, 98
    Delphinus and, 100, 102
    Draco and, 74
    Eridanus and, 120
    Gemini and, 57, 59–60
    Hercules and, 12–13, 78–79, 85
    Leo and, 60, 62
    Lepus and, 123
    Libra and, 155–156
    Lupus and, 158, 160
    Lyra and, 106, 108
    Ophiuchus and, 161, 163
    Orion and, 32, 34
    Pegasus and, 109, 112
    Pelops and, 25
    Perseus and, 35–36
    Phineus and, 131
    Pisces and, 42
    powers of, 4
    Prometheus and, 115
    Roman name for, 223
    Sagitta and, 114
    Sagittarius and, 187, 189
    Scorpius and, 164, 166
    Taurus and, 43, 45
    Triangulum and, 46, 48
    Ursa Major and, 63–64
    Ursa Minor and, 86, 88
    Virgo and, 169, 171
Zodiac, constellations of, 11